It is a massy wheel,
Fixed on the summit of the highest mount,
To whose huge spokes ten thousand lesser things
Are mortis'd and adjoin'd; which when it falls;
Each small annexement, petty consequence,
Attends the boist'rous ruin.
                    Shakespeare: *Hamlet* II, iii

DEDICATION
to my sisters Mrs Helen Facey and Mrs Zee Hunter
and the memory of my late brother Brian Norman Ross

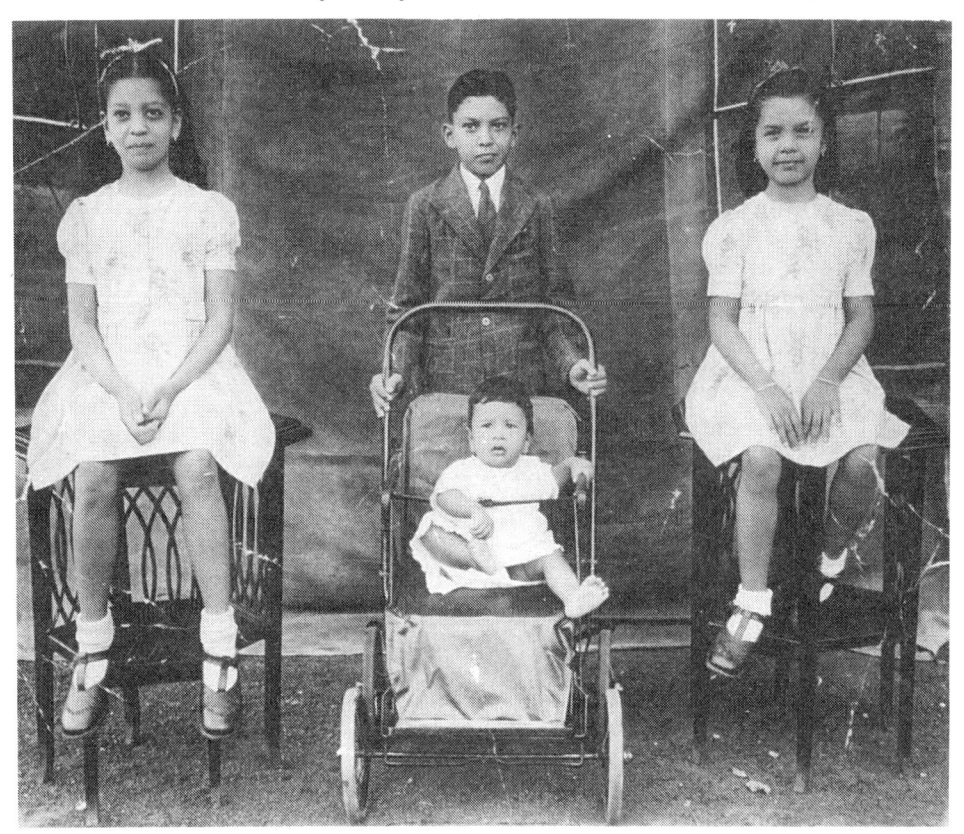

Family photograph, Kharagpur, India, 1948

ABOUT OUR AUTHOR

Carlisle Thomas Francis Ross was born in Kharagpur, India and educated in Bangalore at St Joseph's European High School (1944-47) during the closing years of the British Raj. After coming to England he attended the Chatham Technical School for Boys (1948-51) followed by part time education at the Royal Dockyard Technical College (1951-56) at Chatham, Kent where he served a five-year shipwright apprenticeship.

He proceeded to King's College, Newcastle-upon-Tyne (University of Durham), reading for a B.Sc.(Hons) degree in Naval Architecture (1956-59). During university vacations he worked as a part-time draughtsman at HM Dockyard, Chatham.

For the next two years he laid the foundations of his powerful industrial experience as a Designer in the Project Design Office at Vickers-Armstrongs (Shipbuilders), Barrow-in-Furness (1959-61). His outstanding work there was acknowledged by promotion to the position of Deputy Chief of the Project Design Office. He next worked as a research graduate in the Department of Engineering, University of Manchester (1961-62), where in 1963 he gained his Ph.D. for research in Stress Analysis of Pressure Vessels.

He brought his industrial experience from Vickers-Armstrongs into teaching, first as Lecturer in Civil and Structural Engineering at Constantine College of Technology, Middlesborough (now University of Teesside) (1964-66), and later to Portsmouth Polytechnic (now University of Portsmouth) (1966-71) as Senior Lecturer in Mechanical Engineering, where he still remains as Professor of Structural Mechanics.

His outstanding research in structural mechanics is based on computational methods, tested experimentally with colleagues. He has moreover made important discoveries on the buckling of ring-stiffened cylinders and cones under external pressure, and has also developed the application of microcomputers on finite element analysis. His outstanding contributions to engineering science were recognised in 1992 by the award of D.Sc. for research on Stress Analysis and Structural Dynamics by the CNAA, London.

Finite Element Programs in Structural Engineering and Continuum Mechanics (Albion Publishing Limited, 1996), also by Professor Ross, is the companion book. It contains 24 powerful finite element programs in QUICK BASIC, and covers analysis of problems by computer-aided design and manufacture in mechanical, civil, aeronautical, marine, and electrical engineering, including those connected with heat transfer and acoustic vibrations.

FINITE ELEMENT TECHNIQUES IN STRUCTURAL MECHANICS

Carl T.F. Ross, BSc, PhD, DSc, CEng, FRINA, MSNAME
Professor of Structural Dynamics
Department of Mechanical and Manufacturing Engineering
University of Portsmouth

Albion Publishing
Chichester

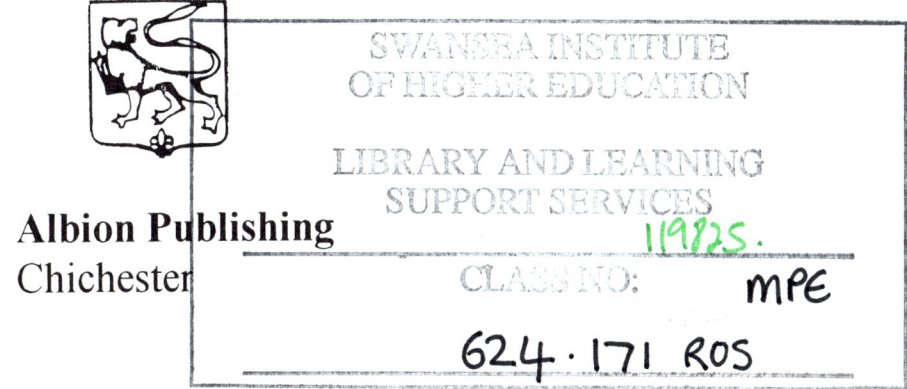

First published in 1996 by
ALBION PUBLISHING LIMITED
International Publishers
Coll House, Westergate, Chichester, West Sussex, PO20 6QL England

COPYRIGHT NOTICE
All Rights Reserved. No part of this publication may be reproduced, stored in a retrieval system, or transmitted, in any form or by any means, electronic, mechanical, photocopying, recording, or otherwise, without the permission of Albion Publishing, International Publishers, Coll House, Westergate, Chichester, West Sussex, England

© Carl T.F. Ross, 1996

British Library Cataloguing in Publication Data
A catalogue record of this book is available from the British Library

ISBN 1-898563- 25-X

Printed in Great Britain by Hartnolls, Bodmin, Cornwall

Table of Contents

		Page
Author's Preface		vii
Notation		viii
Introduction		1

Chapter 1. Introduction to Matrix Algebra — 5
- 1.1 Introduction — 5
- 1.2 Definitions — 5
- 1.3 Some Special Types of Square Matrix — 6
- 1.4 Addition and Subtraction of Matrices — 9
- 1.5 Matrix Multiplication — 10
- 1.6 Matrix Integration and Differentiation — 11
- 1.7 Determinants — 12
- 1.8 Inverse of a Matrix — 14
- 1.9 Solution of Simultaneous Equations — 16
- Examples for Practice 1 — 24

Chapter 2. The Matrix Displacement Method — 26
- 2.1 Introduction — 26
- 2.2 Stiffness Matrix of a Rod Element [k] — 26
- 2.3 Plane Pin-Jointed Trusses — 38
- 2.4 Three Dimensional Trusses — 45
- 2.5 Continuous Beams — 55
- 2.6 Rigid-Jointed Plane Frames — 67
- 2.7 Stiffness Matrix for a Torque Bar — 76
- Examples for Practice 2 — 83

Chapter 3. The Finite Element Method — 88
- 3.1 Introduction — 88
- 3.2 Stiffness Matrix for the in-Plane Triangular Plate — 88
- 3.3 Stiffness for an in-Plane Annular Plate — 95
- 3.4 Three Node Rod Element — 99
- 3.5 Distributed Loads — 104
- 3.6 von Mises Stress (σ_{vm}) — 106
- Examples for Practice 3 — 107

Chapter 4. Vibration of Structures — 108
- 4.1 Introduction — 108
- 4.2 The Elemental Mass Matrix — 108
- 4.3 Mass Matrix for a Rod Element — 110
- 4.4 Vibrations of Pin-Jointed Trusses — 115
- 4.5 Continuous Beams — 138
- 4.6 Rigid-Jointed Plane Frames — 148
- Examples for Practice 4 — 155

Table of Contents

Chapter 5. Non-Linear Structural Mechanics — **158**
    5.1   Introduction — 158
    5.2   Geometrical Non-Linearity — 158
    5.3   Material Non-Linearity — 158
    5.4   Combined Geometrical and Material Non-Linearity — 158
    5.5   Problems Involving Geometric Non-Linearity — 159
    5.6   Problems Involving Elastic Buckling — 169
    5.7   Non-Linear Vibrations — 175
    Examples for Practice 5 — 181

Chapter 6. The Modal Method of Analysis — **183**
    6.1   Introduction — 183
    6.2   The Modal Matrix $[\Phi]$ — 183
    6.3   Damping — 193

References — **198**

Answers to Further Problems — **199**
    Examples for Practice 1 — 199
    Examples for Practice 2 — 202
    Examples for Practice 3 — 203
    Examples for Practice 4 — 204
    Examples for Practice 5 — 205

Author's Preface

The book is aimed at undergraduates and postgraduates in mechanical, civil, structural and aeronautical engineering and naval architecture.

The approach used in the book is a step-by-step methodological one, which addresses the many mathematical difficulties experienced by present-day students. The mathematical and physical theories are carefully explained with the aid of many worked examples, so that these theories effectively become quite simple to understand.

Chapter 1 is on elementary matrix algebra, and reveals most of the matrix algebra required to understand the book. Chapter 2 is on the matrix displacement method, and guides the reader from an application to simple springs, through to two and three dimensional pin-jointed trusses, continuous beams and rigid-jointed plane frames. Distributed loads are also taken into account. Chapter 3 introduces the finite element method proper, and shows how it can be applied to in-plane plates and rods. Chapter 4 is on the vibration of structures and shows how the finite element method can be applied to the vibration of two and three dimensional pin-jointed trusses, continuous beams and rigid-jointed plane frames. Chapter 5 is on non-linear structural mechanics and shows how the finite element method can be applied to buckling and non-linear structural problems, including non-linear structural vibrations. A formula is given for the von Mises stress. Chapter 6 is on the modal method of analysis and shows how the finite element method can be applied quite simply, to problems involving damping and forced vibrations. This chapter shows that for many complex problems involving forced vibrations with damping, their solution can be considerably simplified using modal analysis. Most of the chapters have a section on "Examples for Practice", from which the readers can practice their newly acquired skills.

Carl T F Ross
April 16, 1996

Acknowledgements

The Author would like to thank Mrs Joanna Russell and Miss Sharon Snook for the care and devotion they showed in typing this book.

Notation

Unless otherwise stated, the following symbols are adopted:

A	=	cross-sectional area
I	=	second moment of area
I_c	=	mass moment of inertia
J	=	torsional constant
l	=	length
t	=	thickness or time
T	=	torque
M	=	bending moment
n	=	frequency (Hz)
r	=	radius
R_1, R_2	=	radii at nodes 1 and 2 respectively
E	=	elastic modulus
G	=	rigidity modulus
x, y, z	=	co-ordinates (local axes)
x^o, y^o, z^o	=	co-ordinates (global axes)
X, Y, Z	=	forces in x, y and z directions respectively
X^o, Y^o, Z^o	=	forces in x^o, y^o and z^o directions respectively
u, v, w	=	displacements in x, y and z directions respectively
u^o, v^o, w^o	=	displacements in x^o, y^o and z^o directions respectively
α	=	angle
λ	=	eigenvalue
ω	=	radian frequency
ρ	=	density
σ	=	stress
ε	=	strain
τ_{xy}	=	shear stress in the x-y plane
γ_{xy}	=	shear strain in the x-y plane
υ	=	Poisson's ratio
ξ	=	x/l
$[k]$	=	elemental stiffness matrix in local co-ordinates
$[k^o]$	=	elemental stiffness matrix in global co-ordinates
$[k_G]$	=	geometrical stiffness matrix in local co-ordinates
$[k^o_G]$	=	geometrical stiffness matrix in global co-ordinates
$[m]$	=	elemental mass matrix in local co-ordinates
$[m^o]$	=	elemental mass matrix in global co-ordinates
$[K^o]$	=	system stiffness matrix in global co-ordinates
$[K^o_G]$	=	geometrical system stiffness matrix in global co-ordinates
$[M^o]$	=	system mass matrix in global co-ordinates
$\{P_i\}$	=	a vector of internal nodal forces
$\{q^o\}$	=	a vector of external nodal forces in global co-ordinates
$\{U_i\}$	=	a vector of nodal displacements in local co-ordinates
$\{U^o_i\}$	=	a vector of nodal displacements in global co-ordinates

Notation

$[K_{11}]$	=	that part of the system stiffness matrix that corresponds to the 'free' displacements
$[K_{G11}]$	=	that part of the geometrical system stiffness matrix that corresponds to the 'free' displacements
$[M_{11}]$	=	that part of the system mass matrix that corresponds to the 'free' displacements
$[T]$	=	a matrix of directional cosines
$[I]$	=	identity matrix
$[\ ]$	=	a square or rectangular matrix
$\{\ \}$	=	a column vector
$\lfloor\ \rfloor$	=	a row vector
$[0]$	=	a null matrix
$[\phi]$	=	modal matrix

Parts of the Greek alphabet commonly used in mathematics

α	alpha
β	beta
γ	gamma
δ	delta
Δ	delta (capital)
ε	epsilon
ζ	zeta
η	eta
θ	theta
κ	kappa
λ	lambda
μ	mu
ν	nu
ξ	xi
Ξ	xi (capital)
π	pi
σ	sigma
Σ	sigma (capital)
τ	tau
ϕ	phi
χ	chi
ψ	psi
ω	omega
Ω	omega (capital)

The following abbreviations are used:

1.2E6 = 1.2×10^{6}
1.2E-6 = 1.2×10^{-6}

Introduction

The design of most modern structures is based on small deflection elastic theory. That is, the deformation of the structure is assumed to be linearly proportional to the applied load, and does not suffer permanent deformation. For example, if the applied load is increased by a factor of two, the structure is assumed to increase its deformation by a factor of two. This behaviour is the same as that which occurs with a simple elastic spring. Likewise, in the case of a complex structure, the complex structure is assumed to behave as a complex spring. That is, if the structure is subjected to snow loads, it will sag downwards, and if it is subjected to horizontal wind loads, the structure will sway sideways. Many of these deflections may be difficult to observe with the naked eye, but with the aid of many of the measuring devices that are available today, these deformations can be measured.

The importance of the finite element method is that it can mathematically analyse complex structures. It can, of course, also analyse simple structures, but if a satisfactory closed-loop trivial solution exists for a closed-loop trivial problem, then it is better to use the closed-loop trivial solution for that problem, rather than use the finite element method.

The most important feature of the finite element method, is that it readily lends itself to the writing of computer programs, which can analyse a large number of complex problems with complex boundary conditions. This is because it is usually very difficult to solve a partial differential equation which applies over a complex shape with complex boundary conditions. In the case of the finite element method, the process is to divide this complex shape into several simpler shapes, called finite elements, as shown in Figure I.1.

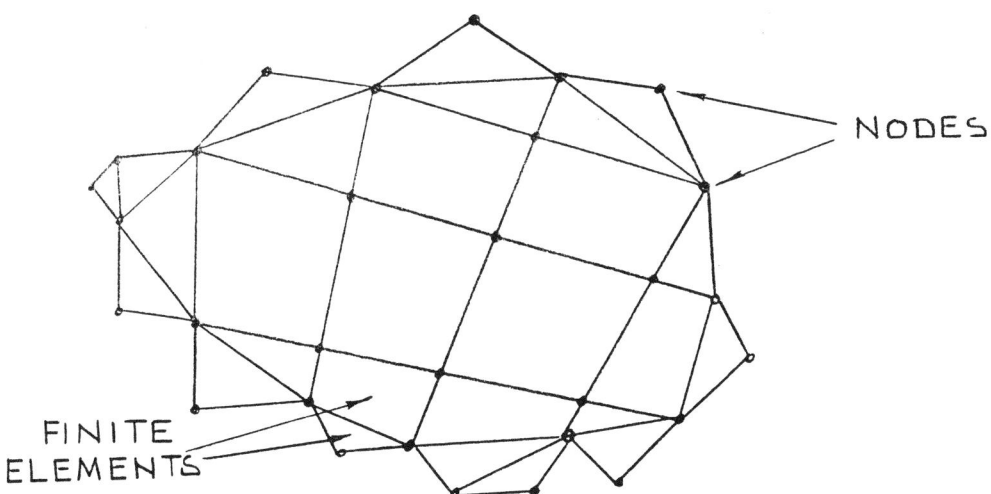

Figure I.1 - Complex shape subdivided into finite elements.

Each finite element, and also the shape of the domain is described by nodes, or nodal points, also shown in Figure I.1.

2 Introduction

The next stage of the process is to solve the partial differential equation over each finite element in turn, and then by considering equilibrium and compatibility at the inter-element boundaries, to join all the finite elements together to obtain the complex parent shape. This process results in a large number of simultaneous equations, the solution of which yields the unknown displacements at the nodes. From these nodal displacements, the stresses in the elements can be determined.

For skeletal structures, such as two and three dimensional, pin-jointed trusses or for two and three dimensional rigid-jointed frameworks, line elements, such as rods or beams can be used, as shown in Figure I.2(a) and (b).

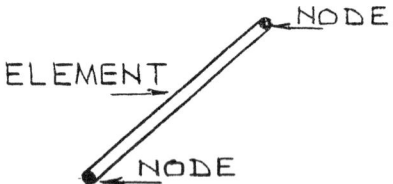

(a) 2 node line element

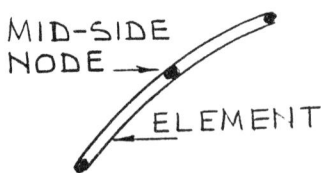

(b) 3 node line element

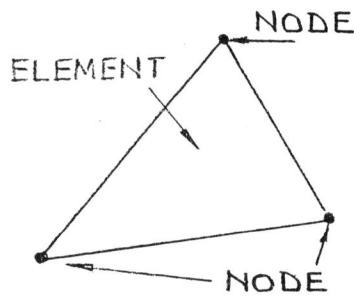

(c) 3 node triangular element

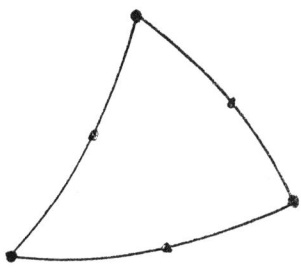

(d) 6 node triangular element

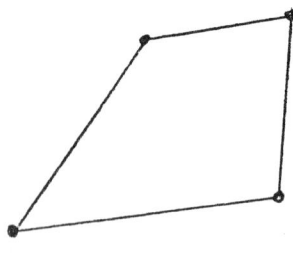

(e) 4 node quadrilateral

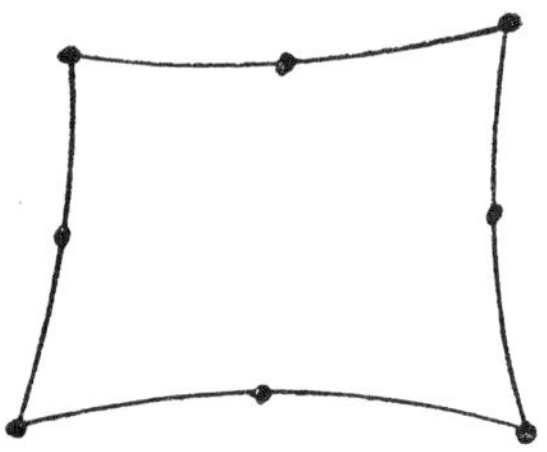

(f) 8 node quadrilateral

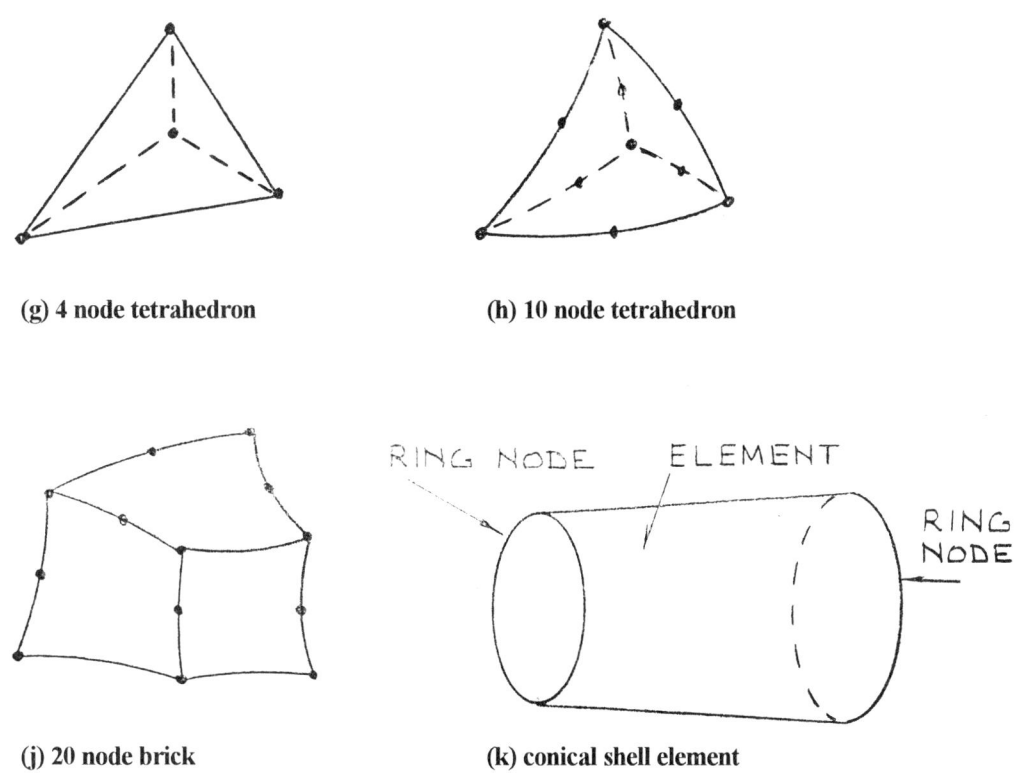

(g) 4 node tetrahedron **(h) 10 node tetrahedron**

(j) 20 node brick **(k) conical shell element**

Figure I.2 - Some typical finite elements.

For plates and shells, two dimensional elements, such as those shown in Figures I.2(c) to (f) can be used, and for solids, block filling elements, such as those shown in Figures I.2 (g) to (j) may prove suitable.

For axisymmetric shells, the truncated conical element for Figure I.2(k) can be used; this element is described by two ring nodes at its ends.

Most finite elements are described by end nodes or corner nodes, but the more reliable elements, usually have mid-side nodes in addition to corner or end nodes.

In fact, the 8 node quadrilateral element of Figure I.2(f) is one of the most popular elements used for analysing in-plane plate problems. It is better known as the 8 node isoparametric element, and to develop it, it is necessary to use Gauss-Legendré numerical integration.

This element, which assumes a parabolic variation for its displacement functions, is more reliable than the 3 and 6 node triangular elements of Figures I.2(c) and I.2(d), and also the four node quadrilateral element of Figure I.2(e).

Similarly, the four node quadrilateral element of Figure I.2(e), which assumes a linear variation for the displacement distributions, is more reliable than the 3 node triangular element of Figure I.2(c). The 6 node triangular element of Figure I.2(d) assumes a parabolic variation for the displacement distributions, but it is not as reliable as the 8 node isoparametric quadrilateral element of Figure I.2(f).

The application of the finite element method to practical problems, is very much dependent on a digital computer, together with a suitable computer program.

The computer program will usually have a **pre-processor**, which will generate the data for the mathematical model required to describe the structure; usually in the form of several hundred simultaneous equations. Using this data file, the computer program will determine the unknown displacements caused by the known externally applied loads. From these nodal displacements, the stresses in the elements will be calculated. Both the solution of the simultaneous equations and the calculations of the stresses will be carried out by a part of the computer program called the **analyser**.

The third part of the computer program is called the **post-processor**. This part of the program uses the values of the calculated nodal displacements and elemental stresses to plot the deformed structure, together with the stress contours. Both the pre-processor and the post-processor are graphical in nature. In addition to having the ability to solve structural problems by the finite element method, the method can solve problems involving vibrations, buckling, shock and impact, acoustics, magnetostatics, electrostatics, fluid flow, heat transfer, etc. However, 95% of finite element applications are to structures.

To avoid numerical instability or round-off error, care should be taken not to join a very flexible element to a very stiff one, or to use triangular or quadrilateral elements which have corner angles larger than 150° or smaller than 30°.

Other numerical techniques used by the engineer are called the finite difference method and the boundary element method. Many of today's researchers on these methods are attempting to produce a unified theory, linking the finite element method to the finite difference method and the boundary element method.

1

Introduction to Matrix Algebra

1.1 Introduction

The approach in this chapter will be one of technique, rather than that of a rigorous mathematical theory. That is, the approach will be one of encouraging the engineer to use matrix algebra as a tool, rather than studying the mathematical wonders of matrix algebra. The finite element method is very much dependent on matrix algebra, and matrix algebra, in turn, lends itself to efficient computer programming.

1.2 Definitions

A **scalar** is a quantity which has magnitude but no direction. Typical scalar quantities include mass time, energy, temperature and speed.

A **vector** is a quantity which has magnitude and direction. Typical vector quantities include force, weight, velocity, acceleration, displacement and torque. For a vector quantity, both the magnitude of the vector and the direction it is acting in are important. It is interesting to note that although energy and torque have the same units, energy is a scalar and torque is a vector. If, however, the torque is multiplied by the angle it turns through, in radians, the resulting units will become scalar.

A **matrix** is a table, whose members are called elements, as shown by equation (1.1).

$$[A] = \begin{bmatrix} a_{11} & a_{12} & a_{13} & - & a_{1n} \\ a_{21} & a_{22} & a_{23} & - & a_{2n} \\ a_{31} & a_{32} & a_{33} & - & a_{3n} \\ a_{m1} & a_{m2} & a_{m3} & - & a_{mn} \end{bmatrix} \tag{1.1}$$

In the matrix of equation (1.1), the elements are a_{11}, a_{12}, a_{13} etc, etc. These elements can be scalars or vectors or even matrices. The matrix of equation (1.1) is said to be a rectangular matrix with m rows and n columns. That is, it is or order m x n.

A **column** is said to be a vertical line of numbers, such as a_{13}, a_{23}, a_{33} ... a_{m3}.
A **row** is said to be a horizontal line of numbers, such as a_{31}, a_{32}, a_{33} ..... a_{3n}.
A typical **column matrix** is shown by equation (1.2).

$$\{A\} = \begin{Bmatrix} a_{11} \\ a_{21} \\ a_{31} \\ - \\ a_{m1} \end{Bmatrix} \tag{1.2}$$

A typical **row matrix** is shown by equation (1.3)

$$[A] = [a_{11} \ a_{12} \ a_{13} \ ... \ a_{1n}] \quad (1.3)$$

Column and row matrices are often called column and row vectors, respectively.

A **square matrix** is one where m = n, as shown by equation (1.4); this matrix is said to be of order n.

$$[A] = \begin{bmatrix} a_{11} & a_{12} & a_{13} & - & a_{1n} \\ a_{21} & a_{22} & a_{23} & - & a_{2n} \\ a_{31} & a_{32} & a_{33} & - & a_{3n} \\ - & - & - & & - \\ a_{n1} & a_{n2} & a_{n3} & - & a_{nn} \end{bmatrix} \quad (1.4)$$

The **transpose** of a matrix $[A]^T$ is obtained by exchanging its rows with its columns.

For example,

$$\text{if } [A] = \begin{bmatrix} 5 & 3 & 1 \\ 2 & 4 & 0 \end{bmatrix}, \text{then } [A]^T = \begin{bmatrix} 5 & 2 \\ 3 & 4 \\ 1 & 0 \end{bmatrix}$$

1.3 Some special types of square matrix

A **diagonal matrix** is one where all the elements are zero, except for the elements on the **leading diagonal**, as shown by equation (1.5).

$$[A] = \begin{bmatrix} a_{11} & 0 & 0 & - & 0 \\ 0 & a_{22} & 0 & - & 0 \\ 0 & 0 & a_{33} & - & 0 \\ - & & & & - \\ 0 & 0 & 0 & - & a_{nn} \end{bmatrix} \quad (1.5)$$

If the elements on the leading diagonal all have the same value, the matrix is said to be a **scalar** matrix, as shown by equation (1.6).

$$[A] = \begin{bmatrix} 7.1 & 0 & 0 & 0 \\ 0 & 7.1 & 0 & 0 \\ 0 & 0 & 7.1 & 0 \\ 0 & 0 & 0 & 7.1 \end{bmatrix} \quad (1.6)$$

If all the elements of the leading diagonal of the scalar matrix are one, the matrix is said to be an **identity** or **unit matrix**, as shown in equation (1.7).

$$[I] = \begin{bmatrix} 1 & 0 & 0 & 0 \\ 0 & 1 & 0 & 0 \\ 0 & 0 & 1 & 0 \\ 0 & 0 & 0 & 1 \end{bmatrix} \quad (1.7)$$

The matrix of equation (1.7) is called a unit matrix, because it is the matrix equivalent of the scalar value of one.

Similarly, the matrix of equation (1.6) is called a scalar matrix, because if it is multiplied into another matrix (say) [B], it increases the value of the elements of the matrix [B], by the value of the scalar quantity in the leading diagonal of the scalar matrix.

An **upper triangular matrix** is a square matrix, which has all its non-zero elements in and above its main diagonal, as shown by equation (1.8).

$$[A] = \begin{bmatrix} a_{11} & a_{12} & a_{13} & - & a_{1n} \\ 0 & a_{22} & a_{23} & - & a_{2n} \\ 0 & 0 & a_{33} & - & a_{3n} \\ | & | & | & & | \\ 0 & 0 & 0 & - & a_{nn} \end{bmatrix} \quad (1.8)$$

A **lower triangular matrix** is a square matrix which has all its non-zero elements in and below its main diagonal, as shown by equation (1.9).

$$[A] = \begin{bmatrix} a_{11} & 0 & 0 & - & 0 \\ a_{21} & a_{22} & 0 & - & 0 \\ a_{31} & a_{32} & a_{33} & - & 0 \\ | & | & | & & | \\ a_{n1} & a_{n2} & a_{n3} & - & a_{nn} \end{bmatrix} \quad (1.9)$$

A **null matrix** is a matrix which has all its elements equal to zero, as shown by equation (1.10) and (1.11).

$$[O_n] = \begin{bmatrix} 0 & 0 & 0 & - & 0 \\ 0 & 0 & 0 & - & 0 \\ 0 & 0 & 0 & - & 0 \\ | & | & | & & | \\ 0 & 0 & 0 & - & 0 \end{bmatrix} \quad (1.10)$$

$$[O_3] = \begin{bmatrix} 0 & 0 & 0 \\ 0 & 0 & 0 \\ 0 & 0 & 0 \end{bmatrix} \quad (1.11)$$

The null matrix of equation (1.10) is of order n and the null matrix of equation (1.11) is of order 3, and this is why these are written $[O_n]$ and $[O_3]$ respectively.

A **band matrix** is one which contains all its non-zero elements within a band about its leading diagonal, as shown by equation (1.12).

$$[A] = \begin{bmatrix} a_{11} & a_{12} & 0 & & & & 0 \\ a_{21} & a_{22} & a_{23} & 0 & & & 0 \\ 0 & a_{32} & a_{33} & a_{34} & 0 & - & 0 \\ 0 & 0 & & & & & | \\ | & | & & & & & 0 \\ 0 & 0 & 0 & - & 0 & a_{n,n-1} & a_{nn} \end{bmatrix} \quad (1.12)$$

The band matrix of equation (1.12) is also known as a tri-diagonal matrix, as the width of its band is three.

A **symmetric matrix** is where all

$a_{ij} = a_{ji}$

A **skew matrix** is where

$a_{ij} = -a_{ji}$ and some $a_{ii} \neq 0$

A **skew symmetric matrix** is where

$a_{ij} = -a_{ji}$ and all $a_{ii} = 0$

The **trace of a matrix** is obtained by summing all the elements on its leading diagonal, as follows:-

$$\text{Trace of } [A] = \sum_{i=1}^{i=n} a_{ii}$$

1.4 Addition and Subtraction of Matrices

$$\text{If } [A] = \begin{bmatrix} 5 & 3 & 1 \\ 2 & 4 & 0 \end{bmatrix}$$

$$\& \ [B] = \begin{bmatrix} -1 & \tfrac{1}{2} & 6 \\ -2 & 7 & 8 \end{bmatrix}$$

$$\text{then } [A] + [B] = \begin{bmatrix} (5-1) & (3+\tfrac{1}{2}) & (1+6) \\ (2-2) & (4+7) & (0+8) \end{bmatrix}$$

$$= \begin{bmatrix} 4 & 3\tfrac{1}{2} & 7 \\ 0 & 11 & 8 \end{bmatrix}$$

$$\& \ [A] - [B] = \begin{bmatrix} (5+1) & (3-\tfrac{1}{2}) & (1-6) \\ (2+2) & (4-7) & (0-8) \end{bmatrix}$$

$$= \begin{bmatrix} 6 & 2\tfrac{1}{2} & -5 \\ 4 & -3 & -8 \end{bmatrix}$$

In general,

$c_{ij} = a_{ij} + b_{ij}$

1.5 Matrix Multiplication

In the matrix equation below

$$[C] = [A] \times [B] \tag{1.13}$$

[A] is known as the **pre-multiplier**
[B] is known as the **post-multiplier**, and
[C] is known as the **product**,

so that, in general

$$c_{ij} = \sum_{i=1}^{m} \sum_{k=1}^{\ell} \sum_{j=1}^{n} a_{ik} \times b_{kj} \tag{1.14}$$

It should be noted from equation (1.14) that the number of columns of [A] should be the same as the number of rows of [B].

To demonstrate matrix multiplication, consider the matrices below.

$$[A] = \begin{bmatrix} 5 & 3 & 1 \\ 2 & 4 & 0 \end{bmatrix}$$

$$[B] = \begin{bmatrix} -1 & 6 \\ -3 & 8 \\ 7 & -2 \end{bmatrix}$$

$$[A][B] = \begin{bmatrix} (5 * -1 + 3 * -3 + 1 * 7) & (5 * 6 + 3 * 8 + 1 * -2) \\ (2 * -1 + 4 * -3 + 0 * 7) & (2 * 6 + 4 * 8 + 0 * -2) \end{bmatrix}$$

$$= \begin{bmatrix} (-5 - 9 + 7) & (30 + 24 - 2) \\ (-2 - 12 + 0) & (12 + 32 - 0) \end{bmatrix}$$

$$= \begin{bmatrix} -7 & 52 \\ -14 & 44 \end{bmatrix}$$

Note, in general

$[A][B] \neq [B][A]$

Some laws of matrix multiplication

$([A][B])[C] = [A]([B][C])$
$[A]([B] + [C]) = [A][B] + [A][C]$
$[A][I] = [I][A]$

If $[A][B] = [C][D]$

then $[B]^T [A]^T = [D]^T [C]^T$

If $[A] = [B][C]$

then $[A]^T = [C]^T [B]^T$

1.6 Matrix integration and differentiation

If $[A] = \begin{bmatrix} 3x & 2x^2 \\ -x^3 & 6 \end{bmatrix}$

$\dfrac{d[A]}{dx} = \begin{bmatrix} 3 & 4x \\ -3x^2 & 0 \end{bmatrix}$

$\& \quad \int_0^1 [A]\, dx = \begin{bmatrix} \dfrac{3x^2}{2} & \dfrac{2x^3}{3} \\ \dfrac{-x^4}{4} & 6x \end{bmatrix}_0^1$

$= \begin{bmatrix} 1.5 & 0.667 \\ -0.25 & 6 \end{bmatrix}$

Some rules in matrix differentiation

$\dfrac{d}{dx}([A] + [B]) = \dfrac{d[A]}{dx} + \dfrac{d[B]}{dx}$

$\dfrac{d}{dx}([A][B]) = \dfrac{d[A]}{dx} \cdot [B] + [A] \dfrac{d[B]}{dx}$

1.7 Determinants

Determinants can only be determined for square matrices; a typical determinant is shown in equation (1.15).

$$|A| = \begin{vmatrix} a_{11} & a_{12} & a_{13} & - & a_{1n} \\ a_{21} & a_{22} & a_{23} & - & a_{2n} \\ a_{31} & a_{32} & a_{33} & - & a_{3n} \\ | & | & | & & | \\ a_{n1} & a_{n2} & a_{n3} & - & a_{nn} \end{vmatrix} \quad (1.15)$$

The rule for expanding second order determinants is as follows:-

$$|A| = \begin{vmatrix} a_{11} & a_{12} \\ a_{21} & a_{22} \end{vmatrix} \quad (1.16)$$

so that

$$\begin{aligned} D &= \text{the value of the determinant} \\ &= a_{11} a_{22} - a_{12} a_{21} \end{aligned} \quad (1.17)$$

The method used for expanding second order determinants can be used for expanding third order determinants as follows:-

$$|A| = \begin{vmatrix} a_{11} & a_{12} & a_{13} \\ a_{21} & a_{22} & a_{23} \\ a_{31} & a_{32} & a_{33} \end{vmatrix} \quad (1.18)$$

so that

$$D = a_{11} \begin{vmatrix} a_{22} & a_{23} \\ a_{32} & a_{33} \end{vmatrix} - a_{12} \begin{vmatrix} a_{21} & a_{21} \\ a_{31} & a_{33} \end{vmatrix} + a_{13} \begin{vmatrix} a_{21} & a_{22} \\ a_{31} & a_{32} \end{vmatrix}$$

$$= a_{11} (a_{22} a_{33} - a_{23} a_{32}) - a_{12} (a_{21} a_{33} - a_{23} a_{31}) + a_{13} (a_{21} a_{32} - a_{22} a_{31}) \quad (1.19)$$

Expansion of determinants of order greater than 3 by the above method is extremely difficult, and other methods have to be used [1].

1.7.1 Minors and Cofactors

A **minor** of a determinant is a smaller determinant, order n-1, which is obtained from the original determinant of order "n", by removing from it equal numbers of columns and rows.

Sec. 1.7] Minors and Cofactors

For example, for the determinant of equation (1.18), some typical minors are:-

$$\begin{vmatrix} a_{22} & a_{23} \\ a_{32} & a_{33} \end{vmatrix} \quad \begin{vmatrix} a_{21} & a_{23} \\ a_{31} & a_{33} \end{vmatrix} \quad \& \quad \begin{vmatrix} a_{21} & a_{22} \\ a_{31} & a_{32} \end{vmatrix}$$

The **cofactor** of the matrix [A] is denoted by $[A]^c$. The cofactors of the third order determinant $|A|$ of equations (1.18) are as follows:-

$$a_{11}^c = \begin{vmatrix} a_{22} & a_{23} \\ a_{32} & a_{33} \end{vmatrix} \qquad a_{12}^c = - \begin{vmatrix} a_{21} & a_{23} \\ a_{31} & a_{33} \end{vmatrix}$$

$$a_{13}^c = \begin{vmatrix} a_{21} & a_{22} \\ a_{31} & a_{32} \end{vmatrix} \qquad a_{21}^c = - \begin{vmatrix} a_{12} & a_{13} \\ a_{32} & a_{33} \end{vmatrix}$$

$$a_{22}^c = \begin{vmatrix} a_{11} & a_{13} \\ a_{31} & a_{33} \end{vmatrix} \qquad a_{23}^c = - \begin{vmatrix} a_{11} & a_{12} \\ a_{31} & a_{32} \end{vmatrix} \qquad (1.20)$$

$$a_{31}^c = \begin{vmatrix} a_{12} & a_{13} \\ a_{22} & a_{23} \end{vmatrix} \qquad a_{32}^c = - \begin{vmatrix} a_{11} & a_{13} \\ a_{21} & a_{23} \end{vmatrix}$$

$$a_{33}^c = \begin{vmatrix} a_{11} & a_{12} \\ a_{21} & a_{21} \end{vmatrix}$$

In general, for an nth order determinant, its cofactors are given by

$$a_{ij}^c = (-1)^{i+j} \begin{vmatrix} a_{11} & a_{12} & \cdots & a_{1,j-1} & a_{1,j+1} & \cdots & a_{1n} \\ a_{21} & a_{22} & \cdots & a_{2,j-1} & a_{2,j+1} & \cdots & a_{2n} \\ \vdots & \vdots & & \vdots & & & \\ a_{i-1,1} & & & & & & \\ a_{i+1,1} & & & & & & \\ \vdots & & & & & & \\ a_{n1} & & & & & & a_{nn} \end{vmatrix} \qquad (1.21)$$

1.7.2 Cofactor Matrix

A cofactor of a matrix $[A]^C$ is obtained by replacing all its elements by their cofactors.

1.7.3 Adjoint or Adjugate Matrix

The **adjoint** of a matrix $[A]^a$ is the transpose of its cofactor.

ie $[A]^a = [A]^{CT}$ (1.22)

1.8 Inverse of a matrix

The **inverse matrix** $[A]^{-1}$ is also known as a **reciprocal matrix**. It is the matrix equivalent of a scalar reciprocal and it is useful when matrix manipulation is required. It can also be used for solving simultaneous equations, but the present author does not recommend it for this purpose.

$$[A^{-1}] = \frac{[A]^a}{\det|A|} \qquad (1.23)$$

That is, the inverse of a matrix, is its adjoint matrix, divided by its determinant.

For the 2 x 2 matrix below

$$[A] = \begin{bmatrix} a_{11} & a_{12} \\ a_{21} & a_{22} \end{bmatrix} \qquad (1.24)$$

and its cofactor matrix is given by

$$[A]^C = \begin{bmatrix} a_{22} & -a_{21} \\ -a_{12} & a_{11} \end{bmatrix} \quad \& \quad [A]^a = \begin{bmatrix} a_{22} & -a_{12} \\ -a_{21} & a_{11} \end{bmatrix}$$

so that its inverse is given by

$$[A]^{-1} = \frac{\begin{bmatrix} a_{22} & -a_{12} \\ -a_{21} & a_{11} \end{bmatrix}}{(a_{11} a_{22} - a_{12} a_{21})} \qquad (1.25)$$

For a 3 x 3 matrix, its inverse can be obtained from equation (1.19) and (1.20).

$$\text{If } [A] = \begin{bmatrix} 3 & 1 & 2 \\ -2 & 4 & -1 \\ 0 & 6 & 5 \end{bmatrix}, \text{ find } [A]^{-1} \qquad (1.26)$$

From equations (1.20)

$$a_{11}^c = \begin{vmatrix} 4 & -1 \\ 6 & 5 \end{vmatrix} = 26$$

$$a_{12}^c = -\begin{vmatrix} -2 & -1 \\ 0 & 5 \end{vmatrix} = 10$$

$$a_{13}^c = \begin{vmatrix} -2 & -4 \\ 0 & 6 \end{vmatrix} = -12$$

$$a_{21}^c = -\begin{vmatrix} 1 & 2 \\ 6 & 5 \end{vmatrix} = 7$$

$$a_{22}^c = \begin{vmatrix} 3 & 2 \\ 0 & 5 \end{vmatrix} = 15$$

$$a_{23}^c = -\begin{vmatrix} 3 & 1 \\ 0 & 6 \end{vmatrix} = -18$$

$$a_{31}^c = \begin{vmatrix} 1 & 2 \\ 4 & -1 \end{vmatrix} = -9$$

$$a_{32}^c = -\begin{vmatrix} 3 & 2 \\ -2 & -1 \end{vmatrix} = -1$$

$$a_{33}^c = \begin{vmatrix} 3 & 1 \\ -2 & 4 \end{vmatrix} = 14$$

From equation (1.19)

$$\begin{aligned} \det [A] &= a_{11}\,a_{11}^c + a_{12}\,a_{12}^c + a_{13}\,a_{13}^c \\ &= 3 \times 26 + 1 \times 10 + 2 \times (-12) \\ &= 64 \end{aligned}$$

Now $[A^{-1}] = \dfrac{[A]^a}{\det[A]}$

$$= \dfrac{\begin{bmatrix} 26 & 7 & -9 \\ 10 & 15 & -1 \\ -12 & -18 & 14 \end{bmatrix}}{64}$$

$$[A]^{-1} = \begin{bmatrix} 0.406 & 0.109 & -0.141 \\ 0.156 & 0.234 & -0.0156 \\ -0.188 & -0.281 & 0.219 \end{bmatrix}$$

(1.27)

1.8.1 Inverse of a diagonal matrix

The inverse of a **diagonal matrix** is as follow:-

$$\begin{bmatrix} a_{11} & 0 & 0 & - & 0 \\ 0 & a_{22} & 0 & - & 0 \\ 0 & 0 & a_{33} & - & 0 \\ | & | & & & | \\ 0 & 0 & 0 & - & a_{nn} \end{bmatrix}^{-1} = \begin{bmatrix} 1/a_{11} & 0 & 0 & - & 0 \\ 0 & 1/a_{22} & & & 0 \\ 0 & 0 & 1/a_{33} & & 0 \\ | & | & & & | \\ 0 & 0 & 0 & - & 1/a_{nn} \end{bmatrix}$$

1.9 Solution of Simultaneous Equations

There are basically two types of simultaneous equations, namely, **non-homogeneous** equations as shown by equation (1.28) and **homogeneous** equations, as shown by equation (1.29). The main difference between these two sets of equations, is that homogeneous equations have zeros on their right.

$a_{11} x_1 + a_{12} x_2 + a_{13} x_3 + - + a_{1n} x_n = c_1$

$a_{21} x_1 + a_{22} x_2 + a_{23} x_3 + - + a_{2n} x_n = c_2$

$a_{n1} x_1 + a_{n2} x_2 + a_{n3} x_3 + - + a_{nn} x_n = c_n$ (1.28)

$a_{11} x_1 + a_{12} x_2 + a_{13} x_3 + - + a_{1n} x_n = 0$

$a_{21} x_1 + a_{22} x_2 + a_{23} x_3 + - + a_{2n} x_n = 0$

$a_{n1} x_1 + a_{n2} x_2 + a_{n3} x_3 + - + a_{nn} x_n = 0$ (1.29)

1.9.1 Solution of non-homogeneous simultaneous equations

Solution of equation (1.28) can be solved in a number of different ways, including by use of the matrix inverse, as shown below.

Equations (1.28) can be written in the matrix form

$$[A] \{x\} = \{C\} \tag{1.30}$$

If [A] is inverted and pre-multiplied into both sides, the following is obtained:-

$$[A]^{-1} [A] \{x\} = [A]^{-1} \{C\} \tag{1.31}$$

or $[I] \{x\} = [A]^{-1} \{C\}$

or $\{x\} = [A]^{-1} \{C\}$ (1.32)

Hence, the unknown vector [x] can be determined from equation (1.32).

Example 1.1

Using equation (1.32), solve the following three simultaneous equations:-

$$3x_1 + x_2 + 2x_3 = 1$$

$$-2x_1 + 4x_2 - x_3 = 2$$
$$6x_2 + 5x_3 = 3 \tag{1.33}$$

$$\text{ie} \begin{bmatrix} 3 & 1 & 2 \\ -2 & 4 & -1 \\ 0 & 6 & 5 \end{bmatrix} \begin{Bmatrix} x_1 \\ x_2 \\ x_3 \end{Bmatrix} = \begin{Bmatrix} 1 \\ 2 \\ 3 \end{Bmatrix}$$

From equation (1.27)

$$\begin{bmatrix} 3 & 1 & 2 \\ -2 & 4 & -1 \\ 0 & 6 & 5 \end{bmatrix}^{-1} = \begin{bmatrix} 0.406 & 0.109 & -0.141 \\ 0.156 & 0.234 & -0.0156 \\ -0.188 & -0.281 & 0.219 \end{bmatrix}$$

$$\text{ie} \begin{Bmatrix} x_1 \\ x_2 \\ x_3 \end{Bmatrix} = \begin{bmatrix} 0.406 & 0.109 & -0.141 \\ 0.156 & 0.234 & -0.0156 \\ -0.188 & -0.281 & 0.219 \end{bmatrix} \begin{Bmatrix} 1 \\ 2 \\ 3 \end{Bmatrix}$$

$$= \begin{Bmatrix} 0.201 \\ 0.577 \\ -0.093 \end{Bmatrix} \quad (1.34)$$

If readers solve equations (1.33) by computer, they will find that there are small errors in the above solution; this is due to round off error and shows the importance of using a sufficient number of significant digits.

However, a simpler way of solving equation (1.33), is to eliminate the equations below the leading diagonal, so that equation (1.28) becomes a **triangular form**, as shown by equation (1.35).

$$
\begin{array}{llll}
a_{11}^1 x_1 + & a_{12}^1 x_2 + & a_{13}^1 x_3 + \ldots\ldots a_{1n}^1 x_n & = c_1^1 \\
 & a_{22}^1 x_2 + & a_{23}^1 x_3 + \ldots + a_{2n}^1 x_n & = c_2^1 \\
 & & a_{33}^1 x_3 + \ldots + a_{3n}^1 x_n & = c_3^1 \\
 & & a_{nn}^1 x_n & = c_n^1
\end{array} \quad (1.35)
$$

Hence, $x_n = c_n^1/a_{nn}^1$, and by back-substitution, other values of the vector [x] can be determined.

Example 1.2
Solve equations (1.33) by the triangulation method.
$$\begin{array}{l} 3x_1 + x_2 + 2x_3 = 1 \\ -2x_1 + 4x_2 - x_3 = 2 \\ 6x_2 + 5x_3 = 3 \end{array} \quad (1.36)$$

Multiply the second line of equation (1.36) by 3/2 to give equations (1.37)
$$\begin{array}{l} 3x_1 + x_2 + 2x_3 = 1 \\ -3x_1 + 6x_2 - 1.5x_3 = 3 \\ 6x_2 + 5x_3 = 3 \end{array} \quad (1.37)$$

From equation (1.37), it can be seen that it is now possible to eliminate the coefficient of x_1 from the second line of this equation. This is done as follows:

Add the first line of equation (1.37) to the second line of equation (1.37), to give the new second line, as shown by equation (1.38); this can be seen to be the process of eliminating the equations below the leading diagonal.

$$\begin{array}{l} 3x_1 + x_2 + 2x_3 = 1 \\ 7x_2 + 0.5x_3 = 4 \\ 6x_2 + 5x_3 = 3 \end{array} \quad (1.38)$$

Multiply the third line of equation (1.38) by 7/6 to give the new third line, as shown by equation (1.39).
$$\begin{array}{l} 3x_1 + x_2 + 2x_3 = 1 \\ 7x_2 + 0.5x_3 = 4 \\ 7x_2 + 5.833x_3 = 3.5 \end{array} \quad (1.39)$$

Solution of Simultaneous Equations

From equation (1.39), it can be seen that it is now possible to eliminate the coefficient of x_2 from the third line of equation (1.39). This is done as follows:

Take away the second line of equation (1.39) from the third line of equation (1.39), to give the new third line, as shown by equation (1.40); this process continues to eliminate the equations below the leading diagonal

$$3x_1 + x_2 + 2x_3 = 1$$
$$7x_2 + 0.5x_3 = 4 \quad (1.40)$$
$$5.333x_3 = -0.5$$

Equations (1.40) can now be seen to be of triangular form.

From the last line of equations (1.40)
$x_3 = -0.5/5.333$
$\underline{x_3 = -0.0938} \quad (1.41)$

Substituting equation (1.41) into the second line of equation (1.40)
$x_2 = (4 - 0.5\, x_3)/7$
$\underline{x_2 = 0.578} \quad (1.42)$

Substituting x_2 and x_3 into the first line of equation (1.40), x_1 can be determined
$x_1 = (1 - x_2 - 2x_3)/3$
$\underline{x_1 = 0.203} \quad (1.43)$

1.9.2 Solution of homogeneous equations

Equations (1.29) can be rewritten in the following matrix from:-

$$[B]\{x\} = \{0\} \quad (1.44)$$

Solution of equation (1.44) results in:
  either $\{x\} = \{0\}$

  or $|B| = 0$

The solution $\{x\} = \{0\}$ is only of mathematical interest, and has no practical significance, hence, we will interest ourselves in the solution:-
$$|B| = 0 \quad (1.45)$$

Solution of equation (1.45) is of importance for determining eigenvalues. This can be shown by considering the matrix equation (1.46)

$$[A]\{x\} = \lambda\{x\} \quad (1.46)$$

where λ = the eigenvalues
Eigenvalues or characteristic values or latent roots, are the roots of a polynomial of order n.

Equation (1.46) can be rewritten in the form

$$([A] - \lambda [I]) \{x\} = \{0\} \tag{1.47}$$

so that
$$|[A] - \lambda [I]| = 0 \tag{1.48}$$

There are several ways of solving equation (1.48) [1,2], but the method used here will be to expand the determinant of equation (1.48), and solve the resulting polynomial. The **eigenmodes** of the matrix are the values of $\{x\}$, corresponding to each eigenvalue.

For example, in the case of vibration, the eigenmodes are the shapes of the modes of vibration of the structure for each particular eigenvalue or resonant frequency.

Example 1.3
Determine the eigenvalues and eigenmodes of the matrix of equation (1.49)

$$\begin{bmatrix} 2 & -1 \\ 1 & 4 \end{bmatrix} \tag{1.49}$$

Equation (1.49) must be rewritten as follows:-

$$\begin{vmatrix} (2 - \lambda) & -1 \\ 1 & (4 - \lambda) \end{vmatrix} = 0 \tag{1.50}$$

Expanding equation (1.5), gives the following second order polynomial:-

$(2 - \lambda)(4 - \lambda) - (-1) \times 1 = 0$

or $8 - 2\lambda - 4\lambda + \lambda^2 + 1 = 0$

or $\lambda^2 - 6\lambda + 9 = 0 \tag{1.51}$

Equation (1.51) is of order 2, because the determinant of equation (1.5) is of order 2.

Solving equation (1.51) by the well-known formula for quadratic equations, the following is obtained for the eigenvalues λ_1 and λ_2.

$$\lambda_1 = \frac{6 \pm \sqrt{[36 - 36]}}{2}$$

$\lambda_1 = 3 \tag{1.52}$

$\lambda_2 = 3 \tag{1.53}$

This problem has equal eigenvalues or equal roots. To determine the eigenmodes, substitute equation (1.52) into the first line of equation (1.46), and equation (1.53) into the second line of equation (1.46) as follows:-

$(2-3)x_1 - 1x_2 = 0$

or $-x_1 - x_2 = 0$

$x_1 = -x_2$

∴ the first eigenmode, corresponding to λ_1 is [1 -1]

For the second eigenmode,
$x_1 + x_2 = 0$

$x_1 = -x_2$

∴ the second eigenmode, corresponding to λ_2 is [-1 1]

Example 1.4
Determine the eigenvalues and eigenmodes for the matrix of equation (1.54).

$$\begin{bmatrix} 2 & -1 \\ -1 & 2 \end{bmatrix} \tag{1.54}$$

$$\therefore \begin{vmatrix} (2 - \lambda) & -1 \\ -1 & (2 - \lambda) \end{vmatrix} = 0 \tag{1.55}$$

or $(2 - \lambda)(2 - \lambda) - 1 = 0$

$4 - 4\lambda + \lambda^2 - 1 = 0$

∴ $\lambda^2 - 4\lambda + 3 = 0$

ie $\lambda = \dfrac{4 \pm \sqrt{(16 - 12)}}{2}$

$= \dfrac{4 \pm 2}{2}$

$\lambda_1 = 1$

$\lambda_2 = 3$

The eigenmode corresponding to λ_1 is obtained by substituting λ_1 into the first line of equation (1.54).

ie $(2 - 1) x_1 - x_2 = 0$

or $x_1 = x_2 = 1$

ie the first eigenmode corresponding to λ_1 is:-

$[x_1 \quad x_2] = [1 \quad 1]$

The second eigenmode can be obtained by substituting λ_2 into the second line of equation (1.55)
ie $\quad -x_1 + (2 - 3) x_2 = 0$

or $x_1 = -x_2$

∴ the second eigenmode corresponding to λ_2 is:-

$[x_1 \quad x_2] = [1 \quad -1]$

It should be noted that the first eigenmode could have been obtained by substituting λ_1 into the second line of equation (1.55), and the second eigenmode could have been obtained by substituting λ_2 into the first line of equation (1.55).

Example 1.5
Determine the eigenvalues and eigenmodes of the matrix of equation (1.56).

$$\begin{bmatrix} -1 & 2 & -1 \\ 0 & -1 & 2 \end{bmatrix} \tag{1.56}$$

or $\begin{vmatrix} -1 & (2 - \lambda) & -1 \\ 0 & -1 & (2 - \lambda) \end{vmatrix} = 0$ \quad (1.57)

or $(2 - \lambda) [(2 - \lambda)(2 - \lambda) - 1] + 1 [(-1).(2 - \lambda) + 0] + 0 = 0$ \quad (1.58)

or

$(2 - \lambda)(4 - 4\lambda + \lambda^2 - 1) - 2 + \lambda = 0$

$(2 - \lambda)(4 - 4\lambda + \lambda^2 - 1 - 1) = 0$

or
$(2 - \lambda)(\lambda^2 - 4\lambda + 2) = 0$

Solution of Simultaneous Equations

One root of which is $\underline{\lambda = 2}$

The other two roots can be obtained from the following quadratic equations:-

$\lambda^2 - 4\lambda + 2 = 0$

$$\lambda = \frac{+4 \pm \sqrt{(16 - 8)}}{2} = \frac{4 \pm 2.828}{2}$$

$\lambda = 0.586 \ \& \ 3.414$

$$\therefore \underline{\lambda_1 = 0.586; \ \lambda_2 = 2; \ \lambda_3 = 3.414}$$

To obtain the first eigenmode, substitute λ_1 into the first and third lines of equation (1.57), to give

$(2 - 0.586) x_1 - x_2 = 0$

$1.414 x_1 - x_2 = 0$

ie $x_2 = 1.414 x_1$ or $x_1 = 0.707 x_2$

and $\quad - x_2 + (2 - 0.586) x_3 = 0$

ie $x_2 = 1.414 x_3$ or $x_3 = 0.707 x_2$

This results in the following eigenmode for λ_1

$[x_1 \ \ x_2 \ \ x_3] = [0.707 \ \ 1 \ \ 0.707]$

To obtain the second eigenmode, corresponding to λ_2, substitute λ_2 into the second line of equation (1.57).

$- x_1 + 0 - x_3 = 0$

$x_1 = - x_3$

Substituting λ_2 into the first line of equation (1.57)

$0 - x_2 = 0$

$x_2 = 0$

$\therefore$ the 2nd eigenmode is
$[x_1 \ \ x_2 \ \ x_3] = [1 \ \ 0 \ \ -1]$

To obtain third eigenmode corresponding to λ_3, substitute λ_3 into the first and third lines of equation (1.57).

$(2 - 3.414) x_1 - x_2 = 0$

$x_1 = -0.707 x_2$

$\& - x_2 + (2 - 3.414) x_3 = 0$

$x_3 = - 0.707 x_2$

$\therefore$ the 3rd eigenmode is

$[x_1 \quad x_2 \quad x_3] = [-0.707 \quad 1 \quad -0.707]$

Examples for Practice 1

1. If $[A] = \begin{bmatrix} 2 & -1 \\ -1 & 2 \end{bmatrix}$

   $\& [B] = \begin{bmatrix} 3 & 1 \\ 2 & 4 \end{bmatrix}$

   (a) Find $[A] + [B]$
   (b) Find $[A] - [B]$
   (c) Find $[B] - [A]$
   (d) Find $[A][B]$
   (e) Find $[B][A]$

2. If $[C] = [1 \quad 2]$

   and $\{D\} = \begin{Bmatrix} 3 \\ 4 \end{Bmatrix}$

   (a) Find $[C]\{D\}$
   (b) Find $\{D\}[C]$

3.
   (a) Find $[A]^{-1}$
   (b) Find $[B]^{-1}$

4. Find the eigenvalues and eigenvectors of $[B]$

5. If $[E] = \begin{bmatrix} 5 & 1 & 3 \\ 2 & 6 & 0 \\ 4 & -1 & 7 \end{bmatrix}$

and $[F] = \begin{bmatrix} 6 & 2 & 1 \\ -1 & 8 & 4 \\ -2 & 3 & 7 \end{bmatrix}$

(a) Find $[E] + [F]$

(b) Find $[E] - [F]$

(c) Find $[E][F]$

(d) Find $[F][E]$

(e) Find $[E]^{-1}$

(f) Find $[F]^{-1}$

(g) Find the eigenvalues and eigenvectors of $[E]$

(h) Find the eigenvalues and eigenvectors of $[F]$

CHAPTER 2

The Matrix Displacement Method

2.1 Introduction

The finite element method is based on the principles of the matrix displacement method. That is, after the stiffness of the structure is obtained in the form of a matrix, the mathematical model of the structure is subjected to a known vector of external loads; this causes the structure to deform into a vector of displacements, which have to be calculated, usually by a digital computer. Once these displacements are known at the nodes of the structure, the stresses in each element can be determined from Hookean elastic theory. The method can be used just as easily for statically indeterminate structures as for statically determinate structures, and discontinuities present no problems.

In this chapter, elemental stiffness matrices will be obtained for rods, beams and torque bars, and application of these elements will be made to the solution of two and three dimensional frameworks, and also to continuous beams.

The chapter will commence with discussing stiffness and obtaining the stiffness matrix of a rod element.

2.2 Stiffness Matrix of a Rod Element [k]

A rod is defined as a structure which can resist loads axially. That is, it only possesses axial stiffness; it has no flexural stiffness. Rods are used as the members of pin-jointed trusses, as shown in Figures 2.1 and 2.2.

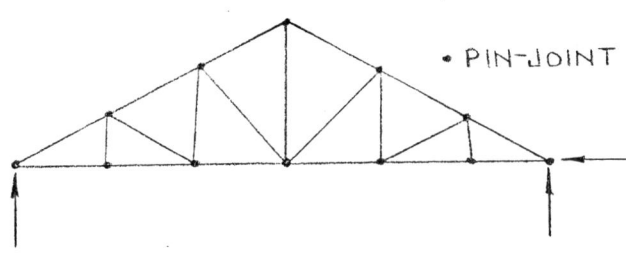

Figure 2.1 - Pin-jointed plane truss.

Sec. 2.2]   **Stiffness Matrix of a Rod Element (k)**   27

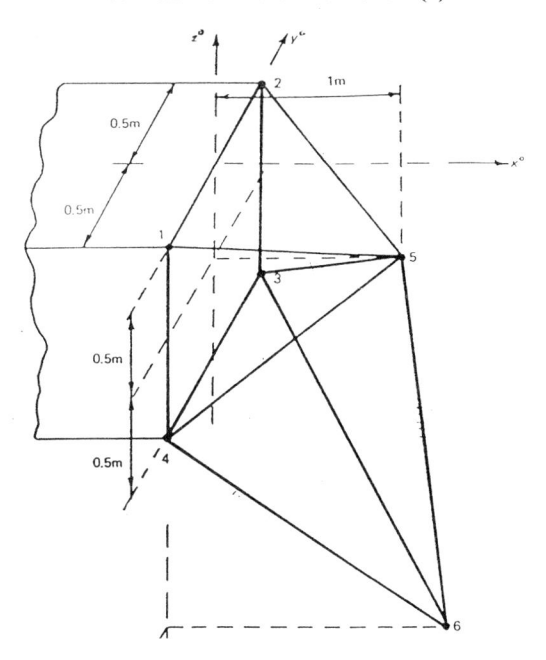

Figure 2.2 - Pin-jointed space truss.

To use the rod element for frameworks, it is important to note that all the joints are pin-jointed, and that all the loads are applied at the joints. If the loads are applied between the joints, or if the joints are rigid (welded), then flexure will occur, and use of the rod element is invalid.

Consider the uniform section rod element of Figure 2.3, which is described by two end nodes, namely node i and node j.

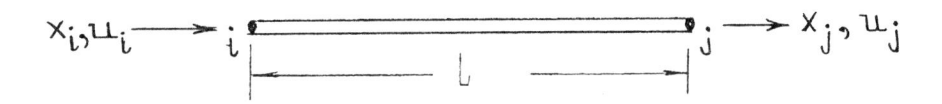

Figure 2.3 - Rod element.

Let,

X_i = axial force at node i

X_j = axial force at node j

u_i = axial deflection at node i

u_j = axial deflection at node j

From Hooke's law,

$$\frac{\text{stress}}{\text{strain}} = E \qquad (2.1)$$

but stress = load/cross-sectional area

$$= X/A \qquad (2.2)$$

and strain = elongation/original length

$$= (u_i - u_j)/\ell \qquad (2.3)$$

where X = axial load

A = cross-sectional area

Substituting equations (2.2) and (2.3) into equation (2.1), the following is obtained:-

$$\frac{X}{A} \cdot \frac{\ell}{(u_i - u_j)} = E$$

or $X = \dfrac{AE}{\ell}(u_i - u_j)$

At node i, $X = X_i$

$$\therefore X_i = \frac{AE}{\ell}(u_i - u_j) \qquad (2.4)$$

From equilibrium considerations,

$X_j = -X_i$

$$\therefore X_j = \frac{AE}{\ell}(u_j - u_i) \qquad (2.5)$$

Rewriting equations (2.4) and (2.5) in matrix form.

$$\begin{Bmatrix} X_i \\ X_j \end{Bmatrix} = \frac{AE}{\ell} \begin{bmatrix} 1 & -1 \\ -1 & 1 \end{bmatrix} \begin{Bmatrix} u_i \\ u_j \end{Bmatrix} \qquad (2.6)$$

or $\{P_i\} = [k]\{U_i\}$ \qquad (2.7)

where

$\{P_i\}$ = a vector of nodal loads = $\begin{Bmatrix} X_i \\ X_j \end{Bmatrix}$ \qquad (2.8)

$[k]$ = elemental stiffness matrix

for a rod = $\dfrac{AE}{\ell} \begin{bmatrix} 1 & -1 \\ -1 & 1 \end{bmatrix}$ \qquad (2.9)

$\{U_i\}$ = a vector of nodal displacements

$$= \begin{Bmatrix} u_i \\ u_j \end{Bmatrix} \qquad (2.10)$$

2.2.1 Structural Stiffness Matrix [K]

In the finite element method, the structure is assumed to consist of a large number of elements and the normal process is to obtain the stiffness matrix of each element in turn and then to mathematically join together these elements to form the stiffness matrix of the entire structure. To demonstrate this process, consider the two element structure of Figure 2.4.

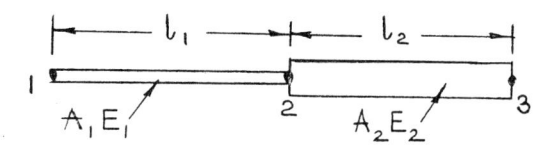

Figure 2.4 - Two element structure.

First let us determine the elemental stiffness matrix of each element in turn.

Element 1-2

Applying equation (2.9) to element 1-2,

$[k_{1\text{-}2}]$ = stiffness matrix of element 1-2

$$[k_{1\text{-}2}] = \frac{A_1 E_1}{\ell_1} \begin{matrix} u_1 & u_2 \\ \begin{bmatrix} 1 & -1 \\ -1 & 1 \end{bmatrix} & \begin{matrix} u_2 \\ u_3 \end{matrix} \end{matrix}$$

(2.11)

Element 2-3

Applying equation (2.9) to element 2-3

$[k_{2\text{-}3}]$ = stiffness matrix of element 2-3

$$[k_{2\text{-}3}] = \frac{A_2 E_2}{\ell_2} \begin{matrix} u_2 & u_3 \\ \begin{bmatrix} 1 & -1 \\ -1 & 1 \end{bmatrix} & \begin{matrix} u_2 \\ u_3 \end{matrix} \end{matrix}$$

(2.12)

The displacements u_1, u_2 and u_3 are not part of the stiffness matrices, but are included to show which part of the elemental stiffness matrices that corresponds to these displacements.

To determine the structural or system stiffness matrix, a matrix of pigeon holes, corresponding to the nodal displacements, u_1, u_2 and u_3, must be drawn, as shown by equation (2.13)

(2.13)

The process now is to join these two elements together, by adding their components of stiffness, corresponding to the structural displacements, u_1, u_2 and u_3, from equation (2.11) and (2.12), into the matrix of pigeon holes, as shown by equation (2.14).

[K] = structural stiffness matrix

$$\begin{bmatrix} & u_1 & u_2 & u_3 & \\ & A_1E_1/\ell_1 & -A_1E_1/\ell_1 & 0 & u_1 \\ & -A_1E_1/\ell_1 & A_1E_1/\ell_1 + A_2E_2/\ell_2 & -A_2E_2/\ell_2 & u_2 \\ & 0 & -A_2E_2/\ell_2 & A_2E_2/\ell_2 & u_2 \end{bmatrix} \qquad (2.14)$$

Example 2.1

A two rod structure is firmly fixed at its top end, and it supports a weight of 10 kN at its bottom end, as shown in Figure 2.5. Determine the resulting nodal deflections.

Element 1-2

$E = 6.7 \times 10^{10}$ N/m^2
$A = 1 \times 10^{-4}$ m^2
$\ell = 1$ m

Element 2-3

$E = 2 \times 10^{11}$ N/m^2
$A = 5 \times 10^{-5}$ m^2
$\ell = 0.5$ m

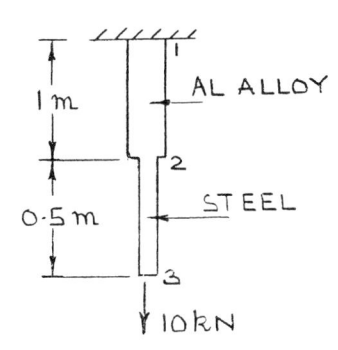

Figure 2.5 - Two element compound bar.

Element 1-2

The stiffness matrix for element 1-2 can be determined from equation (2.9), and as the deflection u_1 = 0, the component of stiffness corresponding to u_1 can be removed, as shown by equation (2.15).

$$[k_{1\text{-}2}] = \begin{bmatrix} u_1 & u_2 \\ & \\ & 6.7 \times 10^6 \end{bmatrix} \begin{matrix} u_1 \\ u_2 \end{matrix} \qquad (2.15)$$

Element 2-3

The stiffness matrix for element 2-3 can be determined from equation (2.9), as shown by equation (2.16).

$$[k_{2\text{-}3}] = \begin{bmatrix} u_2 & u_3 \\ 20 \times 10^6 & -20 \times 10^6 \\ -20 \times 10^6 & 20 \times 10^6 \end{bmatrix} \begin{matrix} u_2 \\ u_3 \end{matrix} \qquad (2.16)$$

The system or structural stiffness matrix is obtained by adding together, the components of stiffness, corresponding to the free displacements, namely, u_2 and u_3, from equations (2.15) and (2.16) as shown by equation (2.17)

$$[k] = \begin{bmatrix} u_2 & u_3 \\ 6.7 \times 10^6 + 20 \times 10^6 & -20 \times 10^6 \\ -20 \times 10^6 & 20 \times 10^6 \end{bmatrix} \begin{matrix} u_2 \\ u_3 \end{matrix} \qquad (2.17)$$

$$= \begin{bmatrix} u_2 & u_3 \\ 26.7 \times 10^6 & -20 \times 10^6 \\ -20 \times 10^6 & 20 \times 10^6 \end{bmatrix} \begin{matrix} u_2 \\ u_3 \end{matrix} \qquad (2.18)$$

The vector of external loads, $\{q_F\}$ corresponding to the free displacements, namely, u_2 and u_3, is shown by equation (2.19)

$$\{q_F\} = \begin{Bmatrix} 0 \\ 10 \times 10^3 \end{Bmatrix} \begin{matrix} u_2 \\ u_3 \end{matrix} \qquad (2.19)$$

The vector of external loads $\{q_F\}$ is obtained by making the first nodal load equal to the value of the nodal load in the direction of the first free displacement, namely u_2; this is zero in this case. The second external load is obtained by making this load equation to the nodal load that corresponds to the second free displacement, namely u_3; this is 10 kN in this case.

Hence, from equation (2.18) and (2.19), the nodal load - nodal displacement relationship for the entire structure is given by:-

$$\left\{\begin{array}{c} 0 \\ 10000 \end{array}\right\} = \begin{bmatrix} 26.7 \times 10^6 & -20 \times 10^6 \\ -20 \times 10^6 & 20 \times 10^6 \end{bmatrix} \left\{\begin{array}{c} u_2 \\ u_3 \end{array}\right\}$$

or $\left\{\begin{array}{c} u_2 \\ u_3 \end{array}\right\} = \begin{bmatrix} 26.7 \times 10^6 & -20 \times 10^6 \\ -20 \times 10^6 & 20 \times 10^6 \end{bmatrix}^{-1} \left\{\begin{array}{c} 0 \\ 10000 \end{array}\right\}$ \hfill (2.20)

From equation (1.25)

$$\begin{bmatrix} 26.7 \times 10^6 & -20 \times 10^6 \\ -20 \times 10^6 & 20 \times 10^6 \end{bmatrix}^{-1} = \frac{\begin{bmatrix} 20 \times 10^6 & 20 \times 10^6 \\ 20 \times 10^6 & 26.7 \times 10^6 \end{bmatrix}}{26.7 \times 10^6 \times 20 \times 10^6 - (20 \times 10^6)^2}$$

$$= \begin{bmatrix} 1.493 \times 10^{-7} & 1.493 \times 10^{-7} \\ 1.493 \times 10^{-7} & 1.993 \times 10^{-7} \end{bmatrix} \hfill (2.21)$$

Substituting equation (2.21) into equation (2.20)

$$\left\{\begin{array}{c} u_2 \\ u_3 \end{array}\right\} = \begin{bmatrix} 1.493 \times 10^{-7} & 1.493 \times 10^{-7} \\ 1.493 \times 10^{-7} & 1.993 \times 10^{-7} \end{bmatrix} \left\{\begin{array}{c} 0 \\ 10000 \end{array}\right\}$$

$$= \left\{\begin{array}{c} 1.493 \times 10^{-3} \text{ m} \\ 1.993 \times 10^{-3} \text{ m} \end{array}\right\}$$

$$\left\{\begin{array}{c} u_2 \\ u_3 \end{array}\right\} = \left\{\begin{array}{c} 1.493 \text{ mm} \\ 1.993 \text{ mm} \end{array}\right\}$$

2.2.2 Effect of settlement or constrained or prescribed displacements

The load-displacement relationship is given by:-

$$\{q\} = [K] \{U\} \tag{2.22}$$

If equation (2.22) is partitioned so that the vector of prescribed displacements $\{u_c\}$, together with the vector of "reactions" $\{R\}$, corresponding to these displacements, is shown by equation (2.23).

$$\left\{\frac{q_F}{R}\right\} = \left[\begin{array}{c|c} K_{11} & K_{12} \\ \hline K_{21} & K_{22} \end{array}\right] \left\{\frac{u_F}{u_c}\right\} \tag{2.23}$$

where,

$\{q\} =$ a vector of external loads

$\{q_F\} =$ a vector of known loads, corresponding to the free degrees of freedom.

$\{R\} =$ a vector of unknown reacting forces, corresponding to the constrained degrees of freedom.

$\{u_F\} =$ a vector of unknown displacements, corresponding to the free degrees of freedom.

$\{u_c\} =$ a vector of known nodal displacements, corresponding to the constrained degrees of freedom.

$[K_{11}] =$ that part of the system stiffness matrix which corresponds to the free degrees of freedom.

Expanding the top row of equation (2.23)

$$\{q_F\} = [K_{11}] \{u_F\} + [K_{12}] \{u_c\} \tag{2.24}$$

or

$$\{u_F\} = [K_{11}]^{-1} (\{q_F\} - [K_{12}] \{u_c\}) \tag{2.25}$$

Substituting equation (2.25) into the second row of equation (2.23), the vector of reactions can be determined from equation (2.26).

$$\{R\} = [K_{21}] \{u_F\} + [K_{22}] \{u_c\} \tag{2.26}$$

if $\{u_c\} = \{0\}$, then

$$\{u_F\} = [K_{11}]^{-1} \{q_F\} \tag{2.27}$$

2.2.3 Stiffness matrix for a rod in global co-ordinates

The limitations of using the stiffness matrix of equation (2.9) is that it can only be applied to a structure where all the rods are in the same straight line. For most plane pin-jointed trusses,

Sec. 2.2] **Stiffness Matrix of a Rod Element [k]** 35

most of the rod elements will be at different angles, as shown in Figure 2.1, and therefore it will be necessary to obtain a general expression for a rod element in two dimensions. Consider the rod element of Figure 2.6, which lies at an angle α to the horizontal axis.

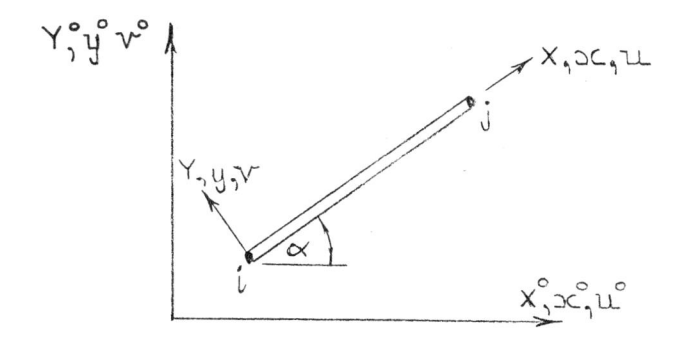

Figure 2.6 - Rod element in global axes.

In Figure 2.6, the global axes are x^o and y^o, and the forces acting in these directions are X^o and Y^o, respectively. The displacements corresponding to x^o and y^o are u^o and v^o, respectively. Similarly, x and y are the local axes, and X and Y are the forces corresponding to these directions. Also u and v are the local displacements corresponding to x and y, respectively.

Now, as the rod only has axial stiffness, $Y_i = Y_j = 0$, and

$$X_i = \frac{AE}{\ell}(u_i - u_j)$$

$$\& \; X_j = \frac{AE}{\ell}(u_j - u_i)$$

or

$$\begin{Bmatrix} X_i \\ Y_i \\ X_j \\ Y_j \end{Bmatrix} = \frac{AE}{\ell} \begin{bmatrix} 1 & 0 & -1 & 0 \\ 0 & 0 & 0 & 0 \\ -1 & 0 & 1 & 0 \\ 0 & 0 & 0 & 0 \end{bmatrix} \begin{Bmatrix} u_i \\ v_i \\ u_j \\ v_j \end{Bmatrix} \quad (2.28)$$

Now, by resolution, it can be seen from Figure 2.6, that

$$X_i = X_i^o \cos \alpha + Y_i^o \sin \alpha$$
$$\& \; Y_i = -X_i^o \sin \alpha + Y_i^o \cos \alpha \quad (2.29)$$

or in matrix form

$$\left\{ \begin{array}{c} X_i \\ Y_i \end{array} \right\} = \left[\begin{array}{cc} c & s \\ -s & c \end{array} \right] \left\{ \begin{array}{c} X_i^\circ \\ Y_i^\circ \end{array} \right\} \qquad (2.30)$$

where

$$\begin{array}{l} c = \cos \alpha = (x_j^\circ - x_i^\circ)/\ell \\ s = \sin \alpha = (y_j^\circ - y_i^\circ)/\ell \\ \ell = \sqrt{[(x_j^\circ - x_i^\circ)^2 + (y_j^\circ - y_i^\circ)^2]} \end{array} \qquad (2.31)$$

$$\text{or} \quad \left\{ \begin{array}{c} X_i \\ Y_i \end{array} \right\} = [T] \left\{ \begin{array}{c} X_i^\circ \\ Y_i^\circ \end{array} \right\} \qquad (2.32)$$

where $\quad [T] = \left[\begin{array}{cc} c & s \\ -s & c \end{array} \right] = $ a coordinate transformation matrix

$$\text{Now } [T]^{-1} = \frac{\left[\begin{array}{cc} c & -s \\ s & c \end{array} \right]}{c^2 + s^2} = \left[\begin{array}{cc} c & -s \\ s & c \end{array} \right]$$

$$\text{or } [T^{-1}] = [T]^T \qquad (2.33)$$

ie The matrix is said to be **orthogonal**, or its inverse is equal to its transpose.

Hence, from equation (2.32)

$$\left\{ \begin{array}{c} X_i^\circ \\ Y_i^\circ \end{array} \right\} = [T]^{-1} \left\{ \begin{array}{c} X_i \\ Y_i \end{array} \right\} = [T]^T \left\{ \begin{array}{c} X_i \\ Y_i \end{array} \right\}$$

Similarly

$$\left\{ \begin{array}{c} u_i \\ v_i \end{array} \right\} = [T] \left\{ \begin{array}{c} u_i^\circ \\ v_i^\circ \end{array} \right\} \qquad (2.34)$$

$$\& \quad \left\{ \begin{array}{c} u_i^\circ \\ v_i^\circ \end{array} \right\} = [T]^T \left\{ \begin{array}{c} u_i \\ v_i \end{array} \right\} \qquad (2.35)$$

Sec. 2.2]   **Stiffness Matrix of a Rod Element [k]**

Hence,

$$\begin{Bmatrix} X_i \\ Y_i \\ X_j \\ Y_j \end{Bmatrix} = \begin{bmatrix} T & O_2 \\ O_2 & T \end{bmatrix} \begin{Bmatrix} X_i^o \\ Y_i^o \\ X_j^o \\ Y_j^o \end{Bmatrix} \qquad (2.36)$$

&

$$\begin{Bmatrix} u_i \\ v_i \\ u_j \\ v_j \end{Bmatrix} = \begin{bmatrix} T & O_2 \\ O_2 & T \end{bmatrix} \begin{Bmatrix} u_i^o \\ v_i^o \\ u_j^o \\ v_j^o \end{Bmatrix} \qquad (2.37)$$

To obtain the load-displacement relationship for a plane rod in global co-ordinates, substitute equations (2.36) and (2.37) into equation (2.28), as follows:

$$\begin{bmatrix} T & O_2 \\ O_2 & T \end{bmatrix} \begin{Bmatrix} X_i^o \\ Y_i^o \\ X_j^o \\ Y_j^o \end{Bmatrix} = \frac{AE}{\ell} \begin{bmatrix} 1 & 0 & -1 & 0 \\ 0 & 0 & 0 & 0 \\ -1 & 0 & 1 & 0 \\ 0 & 0 & 0 & 0 \end{bmatrix} \begin{bmatrix} T & O_2 \\ O_2 & T \end{bmatrix} \begin{Bmatrix} u_i^o \\ v_i^o \\ u_j^o \\ v_j^o \end{Bmatrix}$$

or

$$\begin{Bmatrix} X_i^o \\ Y_i^o \\ X_j^o \\ Y_j^o \end{Bmatrix} = \begin{bmatrix} T & O_2 \\ O_2 & T \end{bmatrix}^T \frac{AE}{\ell} \begin{bmatrix} 1 & 0 & -1 & 0 \\ 0 & 0 & 0 & 0 \\ -1 & 0 & 1 & 0 \\ 0 & 0 & 0 & 0 \end{bmatrix} \begin{bmatrix} T & O_2 \\ O_2 & T \end{bmatrix} \begin{Bmatrix} u_i^o \\ v_i^o \\ u_j^o \\ v_j^o \end{Bmatrix} \qquad (2.38)$$

$$= \frac{AE}{\ell} \begin{bmatrix} a & -a \\ -a & a \end{bmatrix} \begin{Bmatrix} u_i^o \\ v_i^o \\ u_j^o \\ v_j^o \end{Bmatrix} \qquad (2.39)$$

or $\{P°\} = [k°]\{U_i°\}$ (2.40)

where

$$\{P°\} = \begin{Bmatrix} X_i° \\ Y_i° \\ X_j° \\ Y_j° \end{Bmatrix}$$ (2.41)

$$\{U_i°\} = \begin{Bmatrix} u_i° \\ v_i° \\ u_j° \\ v_j° \end{Bmatrix}$$ (2.42)

$$[k°] = \begin{bmatrix} T & O \\ \hline O & T \end{bmatrix}^T [k] \begin{bmatrix} T & O \\ O & T \end{bmatrix}$$

= stiffness matrix for a plane rod in global coordinates

$$= \frac{AE}{\ell} \begin{bmatrix} a & -a \\ \hline -a & a \end{bmatrix}$$ (2.43)

$$[a] = \begin{bmatrix} c^2 & cs \\ cs & s^2 \end{bmatrix}$$ (2.44)

2.3 Plane Pin-Jointed Trusses

In this section, application will be made of the previous sections to the static analysis of plane pin-jointed trusses.

Example 2.2

Determine the nodal displacements and elemental forces in the plane pin-jointed truss of Figure 2.7. All members are of constant AE.

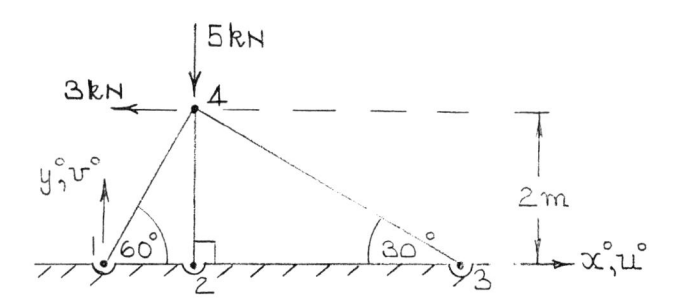

Figure 2.7 Plane pin-jointed truss.

Element 1-4

i = 1, j = 4

This element is said to point from node 1 to node 4, so that its local x axis is as shown in Figure 2.8. That is, the element's start node is 1 and its finish node is 4.

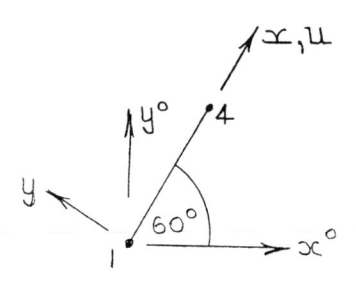

Figure 2.8 Member 1-4.

From Figure 2.8, it can be seen that the angle α, that the local x axes makes with the global x° axis is 60°.

∴ c = cos 60 = 0.5 and s = sin 60 = 0.866

ℓ_{1-4} = 2/sin 60 = 2.309 m

Substituting c, s and ℓ_{1-4} into equation (2.43), and removing the rows and columns corresponding to the zero displacements, which in this case are u_1^o and v_1^o, the elemental stiffness matrix for element 1-4 is given by:-

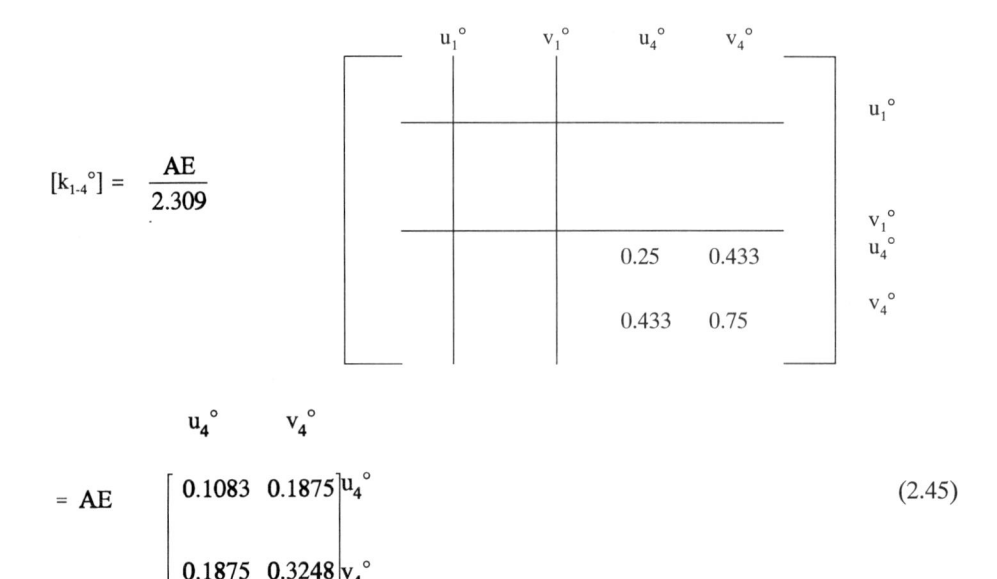

$$[k_{1-4}^o] = \frac{AE}{2.309} \begin{bmatrix} & & & \\ & & & \\ & & 0.25 & 0.433 \\ & & 0.433 & 0.75 \end{bmatrix} \begin{matrix} u_1^o \\ v_1^o \\ u_4^o \\ v_4^o \end{matrix}$$

with columns labelled $u_1^o, v_1^o, u_4^o, v_4^o$

$$= AE \begin{bmatrix} 0.1083 & 0.1875 \\ 0.1875 & 0.3248 \end{bmatrix} \begin{matrix} u_4^o \\ v_4^o \end{matrix} \qquad (2.45)$$

Element 2-4

i = 2 and j = 4

In this case, the element points from node 2 to node 4, so that its local x axis is as shown in Figure 2.9. That is, the member's start node is 2 and its finish node is 4.

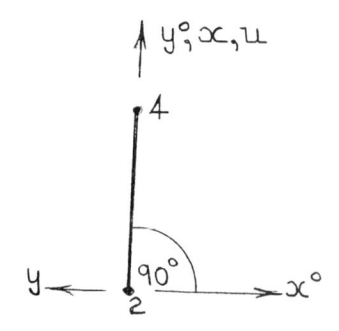

Figure 2.9 Element 2-4.

From Figure 2.9, it can be seen that the angle α, that the local x axis makes with the global x^o axis is 90°.

$\therefore c = \cos 90° = 0$ and $s = \sin 90° = 1$

$\ell_{2-4} = 2$ m

Substituting c, s and ℓ_{2-4} into equation (2.43), and removing the rows and columns corresponding to the zero displacements, which in this case are u_2^o and v_2^o, the elemental stiffness matrix for element 2-4 is given by:-

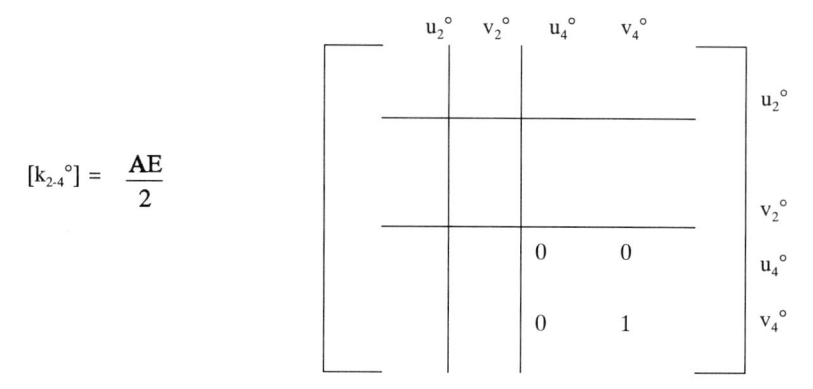

$$[k_{2-4}^o] = \frac{AE}{2} \begin{bmatrix} & u_2^o & v_2^o & u_4^o & v_4^o & \\ & & & & & u_2^o \\ & & & & & \\ & & & & & v_2^o \\ & & & 0 & 0 & u_4^o \\ & & & 0 & 1 & v_4^o \end{bmatrix}$$

$$= AE \begin{bmatrix} u_4^o & v_4^o & \\ 0 & 0 & u_4^o \\ 0 & 0.5 & v_4^o \end{bmatrix} \quad (2.46)$$

Element 4-3

i = 4 and j = 3

In this case, the element points from node 4 to node 3, so that its local x axis is as shown in Figure 2.10. That is, the member's start node is 4 and its finish node is 3.

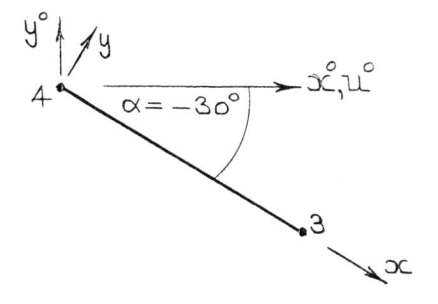

Figure 2.10 Element 4-3.

From Figure 2.10, it can be seen that the local x axis is 30° clockwise from the global x° axis, that is, $\alpha = -30°$.

$\therefore c = \cos(-30°) = 0.866$ and $s = \sin(-30°) = -0.5$ and $\ell_{4-3} = 4m$

Substituting c, s and ℓ_{4-3} into equation (2.43), and removing the columns and rows corresponding to the zero displacements, which in this case is $u_3°$ and $v_3°$, the elemental stiffness matrix for element 4-3 is given by:-

$$[k_{4-3}°] = \frac{AE}{4} \begin{bmatrix} & u_4° & v_4° & u_3° & v_3° \\ & 0.75 & -0.433 & & \\ & -0.433 & 0.25 & & \\ & & & & \\ & & & & \end{bmatrix} \begin{matrix} u_4° \\ v_4° \\ u_3° \\ v_3° \end{matrix}$$

$$= AE \begin{bmatrix} u_4° & v_4° \\ 0.1875 & -0.1083 \\ -0.1083 & 0.0625 \end{bmatrix} \begin{matrix} u_4° \\ v_4° \end{matrix} \qquad (2.47)$$

To obtain that part of the system stiffness matrix, $[K_{11}]$ corresponding to the free displacements, namely $u_4°$ and $v_4°$, the coefficients of the three elemental stiffness matrices, namely equations (2.45) to (2.47) corresponding to these displacements, must be added together, as shown by equation (2.48).

$$[K_{11}] = AE \begin{bmatrix} u_4° & v_4° \\ \begin{array}{c} 0.1083 \\ +0 \\ +0.1875 \end{array} & \begin{array}{c} 0.1875 \\ +0 \\ -0.1083 \end{array} \\ \begin{array}{c} 0.1875 \\ +0 \\ -0.1083 \end{array} & \begin{array}{c} 0.3248 \\ +0.5 \\ +0.0625 \end{array} \end{bmatrix} \begin{matrix} u_4° \\ v_4° \end{matrix} \qquad (2.48)$$

$$= AE \begin{bmatrix} 0.2958 & 0.0792 \\ 0.0792 & 0.8873 \end{bmatrix} \qquad (2.49)$$

Sec. 2.3]   Plane Pin-Jointed Trusses   43

The vector of external loads, namely $\{q_F\}$ are obtained by considering the values of the loads in the direction of the free displacements, namely u_4^o and v_4^o, as follows:-

$$\{q_F\} = \begin{Bmatrix} -3kN \\ -5kN \end{Bmatrix} \begin{matrix} u_4^o \\ v_4^o \end{matrix} \qquad (2.50)$$

Now, from equation (2.27)

$$\{u_F\} = \begin{Bmatrix} u_4^o \\ v_4^o \end{Bmatrix} = [K_{11}]^{-1} \{q_F\}$$

$$= \frac{\dfrac{1}{AE}\begin{bmatrix} 0.8873 & -0.0792 \\ -0.0792 & 0.2958 \end{bmatrix}}{0.2958 \times 0.8873 - 0.0792^2} \begin{Bmatrix} -3 \\ -5 \end{Bmatrix}$$

$$= \frac{1}{AE} \begin{bmatrix} 3.4634 & -0.3091 \\ -0.3091 & 1.1546 \end{bmatrix} \begin{Bmatrix} -3 \\ -5 \end{Bmatrix}$$

or $\begin{Bmatrix} u_4^o \\ v_4^o \end{Bmatrix} = \dfrac{1}{AE} \begin{Bmatrix} -8.845 \\ -4.846 \end{Bmatrix}$ (2.51)

The elemental forces can be obtained by resolving these global displacements along the local axes of each element, and then use Hooke's law, as follows:-

Element 1-4

$i = 1$ and $j = 4$

The local displacement u_4, which acts along the local x axis of element 1-4, at node 4, can be obtained from equation (2.37), as follows:-

$$u_4 = [c \ s] \begin{Bmatrix} u_4° \\ v_4° \end{Bmatrix}$$

$$= [0.5 \ 0.866] \ \frac{1}{AE} \begin{Bmatrix} -8.845 \\ -4.846 \end{Bmatrix}$$

$$y_4 = -8.619/AE$$

Now $u_1 = 0$, as node 1 is securely anchored, hence, from equation (2.4), or Hooke's law, the force in element 1-4 is:

$$F_{1-4} = \frac{AE}{\ell_{1-4}} (u_4 - u_1)$$

$$= \frac{AE}{2.309} \times \left(\frac{-8.619}{AE} - 0 \right)$$

$\underline{F_{1-4} = -3.732 \text{ kN (compressive)}}$

Element 2-4

$i = 2$ and $j = 4$

By using a process similar to that for element 1-4,

u_4 = axial deflection along the local axis x, for element 2-4, at node 4

$$= [c \ s] \begin{Bmatrix} u_4° \\ v_4° \end{Bmatrix}$$

$$u_4 = [0 \ 1] \ \frac{1}{AE} \begin{Bmatrix} -8.845 \\ -4.846 \end{Bmatrix}$$

$$u_4 = -4.846/AE$$

From Hooke's law or equation (2.4), the force in element 2-4 is:

$$F_{2-4} = \frac{AE}{\ell_{2-4}} (u_4 - u_2)$$

$$= \frac{AE}{2} \left(\frac{-4.846}{AE} - 0 \right)$$

$\underline{F_{2-4} = -2.423 \text{ kN (compressive)}}$

Element 4-3

$i = 4$ and $j = 3$

u_4 = axial deflection along the local x axis for element 4-3, at node 4

$$= [c \quad s] \begin{Bmatrix} u_4^\circ \\ v_4^\circ \end{Bmatrix}$$

$$= [0.866 \quad -0.5] \frac{1}{AE} \begin{Bmatrix} -8.845 \\ -4.846 \end{Bmatrix}$$

$\underline{u_4 = -5.237/AE}$

From Hooke's law or equation (2.4), the force in element 4-3 is:

$$F_{4-3} = \frac{AE}{\ell_{4-3}} (u_3 - u_4) = \frac{AE}{4} \left(0 + \frac{5.237}{AE} \right)$$

$\underline{F_{4-3} = 1.309 \text{ kN (tensile)}}$

2.4 Three-dimensional trusses

The three dimensional rod elements is shown in Figure 2.11, where it can be seen that

$$X_i = \text{axial force at node i} = \frac{AE}{\ell} (u_i - u_j) \qquad (2.52)$$

X_j = axial force at node j = $\frac{AE}{\ell} (u_j - u_i)$

$Y_i = Y_j = 0$

$Z_i = Z_j = 0$

X_i° = force in x° direction at node i

X_j° = force in x° direction at node j

Y_i° = force in y° direction at node i $\qquad (2.53)$

Y_j° = force in y° direction at node j

Z_i^o = force in z^o direction at node i

Z_j^o = force in z^o direction at node j

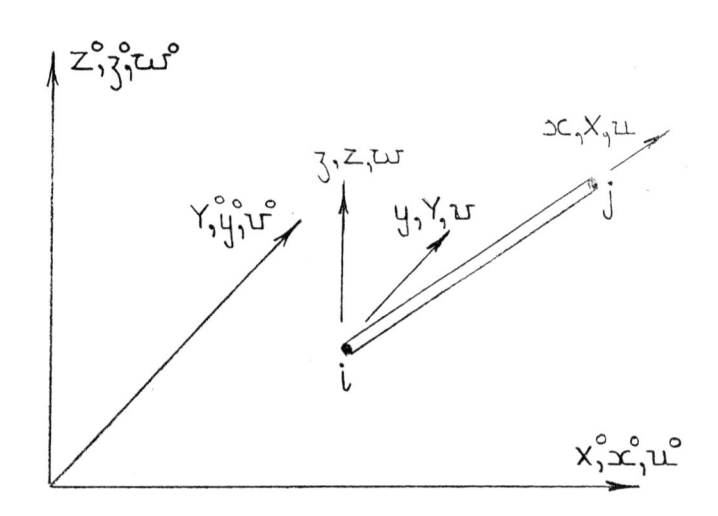

Figure 2.11 Rod in three dimensions.

Resolving along the x axis, it can be shown that

$$X = X^o C_{x,x^o} + Y^o C_{x,y^o} + Z^o C_{x,z^o} \qquad (2.54)$$

where

C_{x,x^o}, C_{x,y^o} and C_{x,z^o} = directional cosines along the x axis

similarly, it can be shown that

$$Y = X^o C_{y,x^o} + Y^o C_{y,y^o} + Z^o C_{y,z^o}$$

$$Z = X^o C_{z,x^o} + Y^o C_{z,y^o} + Z^o C_{z,z^o} \qquad (2.55)$$

or in matrix form

$$\begin{Bmatrix} X \\ Y \\ Z \end{Bmatrix} = \begin{bmatrix} C_{x,x^o} & C_{x,y^o} & C_{x,z^o} \\ C_{y,x^o} & C_{y,y^o} & C_{y,z^o} \\ C_{z,x^o} & C_{z,y^o} & C_{z,z^o} \end{bmatrix} \begin{Bmatrix} X^o \\ Y^o \\ Z^o \end{Bmatrix} \qquad (2.56)$$

$$= [T] \begin{Bmatrix} X^o \\ Y^o \\ Z^o \end{Bmatrix} \qquad (2.57)$$

Similarly,

$$\begin{Bmatrix} u \\ v \\ w \end{Bmatrix} = [T] \begin{Bmatrix} u^\circ \\ v^\circ \\ w^\circ \end{Bmatrix} \quad (2.58)$$

where

[T] = a matrix of directional cosines.

It should be noted that [T] is an <u>orthogonal matrix</u>, so that

$$[T^{-1}] = [T]^T \quad (2.59)$$

Rewriting equation (2.52) in matrix form

$$\begin{Bmatrix} X_i \\ Y_i \\ Z_i \\ X_j \\ Y_j \\ Z_j \end{Bmatrix} = \frac{AE}{\ell} \begin{bmatrix} 1 & 0 & 0 & -1 & 0 & 0 \\ 0 & 0 & 0 & 0 & 0 & 0 \\ 0 & 0 & 0 & 0 & 0 & 0 \\ -1 & 0 & 0 & 1 & 0 & 0 \\ 0 & 0 & 0 & 0 & 0 & 0 \\ 0 & 0 & 0 & 0 & 0 & 0 \end{bmatrix} \begin{Bmatrix} u_i \\ v_i \\ w_i \\ u_j \\ v_j \\ w_j \end{Bmatrix} \quad (2.60)$$

$$= [k]\{U_i\}$$

where

$$[k] = \frac{AE}{\ell} \begin{bmatrix} 1 & 0 & 0 & -1 & 0 & 0 \\ 0 & 0 & 0 & 0 & 0 & 0 \\ 0 & 0 & 0 & 0 & 0 & 0 \\ -1 & 0 & 0 & 1 & 0 & 0 \\ 0 & 0 & 0 & 0 & 0 & 0 \\ 0 & 0 & 0 & 0 & 0 & 0 \end{bmatrix} \quad (2.61)$$

where,

[k] = the elemental stiffness matrix for a rod element in local co-ordinates.

Now from equation (2.38)

[k°] = elemental stiffness matrix for a rod in global coordinates

$$= \begin{bmatrix} T & O_3 \\ O_3 & T \end{bmatrix}^T [k] \begin{bmatrix} T & O_3 \\ O_3 & T \end{bmatrix} \qquad (2.62)$$

$$[k°] = \frac{AE}{\ell} \begin{bmatrix} a & -a \\ -a & a \end{bmatrix} \qquad (2.63)$$

where,

$$[a] = \begin{bmatrix} C_{x,x°}^2 & C_{x,x°}C_{x,y°} & C_{x,x°}C_{x,z°} \\ C_{x,x°}C_{x,y°} & C_{x,y°}^2 & C_{x,y°}C_{x,z°} \\ C_{x,x°}C_{x,z°} & C_{x,y°}C_{x,z°} & C_{x^2,z°} \end{bmatrix} \qquad (2.64)$$

$C_{x,x°} = (x_j° - x_i°)/\ell$

$C_{x,y°} = (y_j° - y_i°)/\ell \qquad (2.65)$

$C_{x,z°} = (z_j° - z_i°)/\ell$

$\ell = \sqrt{[(x_j° - x_i°)^2 + (y_j° - y_i°)^2 + (z_j° - z_i°)^2]} \qquad (2.66)$

where equation (2.66) is Pythagoras' theorem in three dimensions.

Example 2.3

Using the matrix displacement method determine the nodal displacements and elemental forces for the pin-jointed truss of Figure 2.12. This tripod can be assumed to be firmly anchored at nodes 1 to 3.

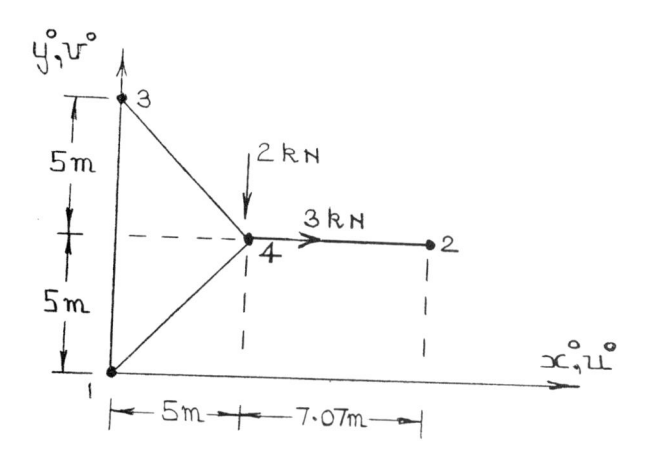

(a) Plan view of tripod

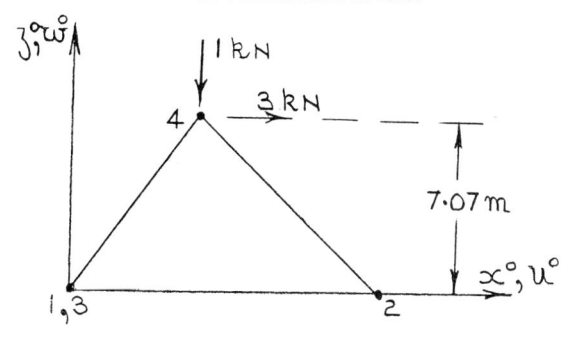

Sec. 2.4] Three-Dimensional Trusses 49

Figure 2.12 Tripod structure.

(b) Front view of tripod

Element 1-4

i = 1 and j = 4

The element points from node 1 to node 4, so that its x axis is in the direction 1 to 4

$x_1^o = y_1^o = z_1^o = 0$

$x_4^o = 5$ m; $y_4^o = 5$m; $z_4^o = 7.07$ m

Substituting the above data into equations (2.65) and (2.66), the following is obtained for the directional cosines and ℓ_{1-4}

$$\ell_{1-4} = \sqrt{[(5-0)^2 + (5-0)^2 + (7.07-0)^2]}$$

$$= 10 \text{ m}$$

$C_{x,x^o} = (5 - 0)/\ell_{1-4} = 0.5$

$C_{x,y^o} = (5 - 0)/\ell_{1-4} = 0.5$

$C_{x,z^o} = (7.07 - 0)/\ell_{1-4} = 0.707$ \hfill (2.67)

Substituting the above data into equation (2.63) and removing the columns and rows corresponding to the zero displacements, which in this case are u_1^o, v_1^o and w_1^o, the following is obtained for the stiffness matrix for element 1-4.

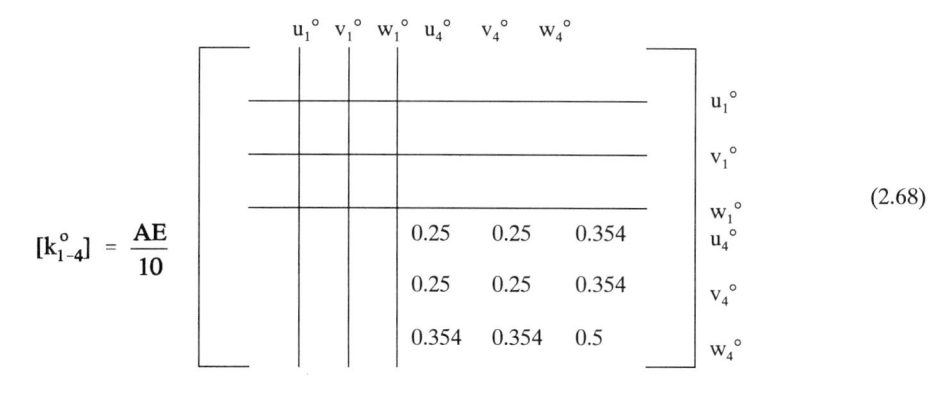

$$[k_{1-4}^o] = \frac{AE}{10} \begin{bmatrix} & & & & & \\ & & & & & \\ & & & & & \\ & & & 0.25 & 0.25 & 0.354 \\ & & & 0.25 & 0.25 & 0.354 \\ & & & 0.354 & 0.354 & 0.5 \end{bmatrix} \begin{matrix} u_1^o \\ v_1^o \\ w_1^o \\ u_4^o \\ v_4^o \\ w_4^o \end{matrix}$$

(2.68)

Element 2-4

i = 2, j = 4

The element points from node 2 to node 4, so that its x axis is along the direction 2 to 4

$x_2° = 12.07$; $y_2° = 5$ m; $z_2° = 0$ m

$x_4° = 5$ m; $y_4 = 5$ m; $z_4° = 7.07$ m

From equation (2.66),

$\ell_{2-4} \sqrt{[(x_4° - x_2°)^2 + (y_4° - y_2°)^2 + (z_4° - z_2°)^2]}$

$= \sqrt{[(5 - 12.07)^2 + (5 - 5)^2 + (7.07 - 0)^2]}$

$\ell_{2-4} = 10$ m

$C_{x,x°} = (5 - 12.07)/10 = -0.707$

$C_{x,y°} = (5 - 5)/10 = 0$ (2.69)

$C_{x,z°} = (7.07 - 0)/10 = 0.707$

Substituting the above data into equation (2.63) and removing the rows and columns corresponding to the zero displacements, which in this case are $u_2°$, $v_2°$ and $w_2°$, the following is obtained for the stiffness matrix of element 2-4.

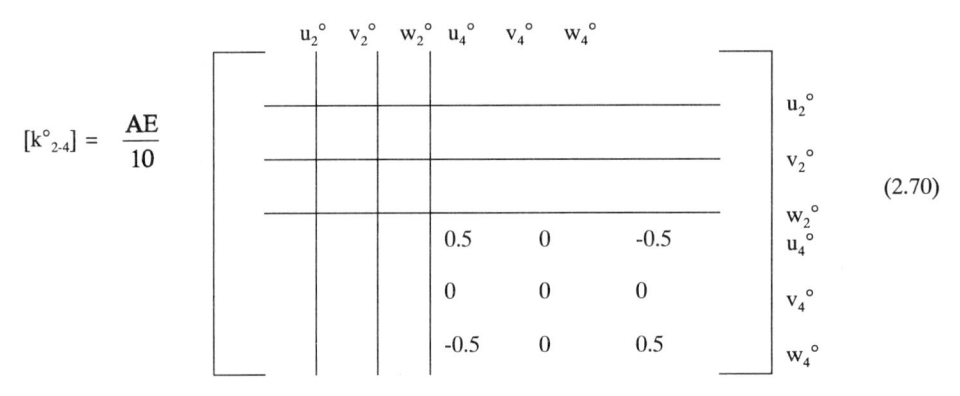

$$[k°_{2-4}] = \frac{AE}{10} \begin{bmatrix} & & & & & & \\ & & & & & & \\ & & & & & & \\ & & & 0.5 & 0 & -0.5 & \\ & & & 0 & 0 & 0 & \\ & & & -0.5 & 0 & 0.5 & \end{bmatrix} \begin{matrix} u_2° \\ v_2° \\ w_2° \\ u_4° \\ v_4° \\ w_4° \end{matrix}$$ (2.70)

with columns labeled $u_2°$, $v_2°$, $w_2°$, $u_4°$, $v_4°$, $w_4°$

Element 4-3

i = 4 and j = 3

The element points from node 4 to node 3 so that its x axis is in the direction from node 4 to node 3.

$x_3° = 0$; $y_3° = 10$ m; $z_3° = 0$

$x_4° = 5$ m; $y_4° = 5$ m; $z_4 = 7.07$ m

Sec. 2.4]   Three-Dimensional Trusses   51

From equation (2.66)

$$\ell_{4-3} = \sqrt{[(x_3^\circ - x_4^\circ)^2 + (y_3^\circ - y_4^\circ)^2 + (z_3 - z_4^\circ)^2]}$$

$$= \sqrt{[(0 - 5)^2 + (10 - 5)^2 + (0 - 7.07)^2]}$$

$$= \sqrt{[25 + 25 + 50]} = 10 \text{ m}$$

$$C_{x,x^\circ} = (0 - 5)/10 = -0.5$$

$$C_{x,y^\circ} = (10 - 5)/10 = 0.5 \qquad (2.71)$$

$$C_{x,y^\circ} = (0 - 7.07)/10 = -0.707$$

Substituting the above data into equation (2.63), and removing the columns and rows corresponding to the zero displacements, which in this case are u_3°, v_3° and w_3°, the following is obtained for the elemental stiffness matrix for element 4-3.

$$[k^\circ_{4-3}] = \frac{AE}{10} \begin{bmatrix} & u_4^\circ & v_4^\circ & w_2^\circ & u_3^\circ & v_3^\circ & w_3^\circ \\ 0.25 & -0.25 & 0.354 & & & \\ -0.25 & 0.25 & -0.354 & & & \\ 0.354 & -0.354 & 0.5 & & & \\ \hline & & & & & \\ \hline & & & & & \\ \hline & & & & & \end{bmatrix} \begin{matrix} u_4^\circ \\ v_4^\circ \\ w_4^\circ \\ u_3^\circ \\ v_3^\circ \\ w_3^\circ \end{matrix} \qquad (2.72)$$

To obtain that part of the system stiffness matrix $[K_{11}]$, which corresponds to the free displacements, namely, u_4°, v_4° and w_4°, the coefficients of the elemental stiffness matrices of equations (2.68), (2.70) and (2.72) should be added together, taking into consideration the appropriate combinations of these free displacements in obtaining $[K_{11}]$.

$$[K_{11}] = \frac{AE}{10} \begin{bmatrix} u_4^\circ & v_4^\circ & w_4^\circ \\ \begin{matrix}0.25 \\ +0.5 \\ +0.25\end{matrix} & \begin{matrix}0.25 \\ +0 \\ -0.25\end{matrix} & \begin{matrix}0.354 \\ -0.5 \\ +0.354\end{matrix} \\ \begin{matrix}0.25 \\ +0 \\ -0.25\end{matrix} & \begin{matrix}0.25 \\ +0 \\ +0.25\end{matrix} & \begin{matrix}0.354 \\ +0 \\ -0.354\end{matrix} \\ \begin{matrix}0.354 \\ -0.5 \\ +0.354\end{matrix} & \begin{matrix}0.354 \\ +0 \\ -0.354\end{matrix} & \begin{matrix}0.5 \\ +0.5 \\ +0.5\end{matrix} \end{bmatrix} \begin{matrix} u_4^\circ \\ \\ v_4^\circ \\ \\ w_4^\circ \end{matrix}$$

$$= \frac{AE}{10}\begin{array}{c} \\ \\ \\ \\ \end{array}\begin{array}{ccc} u_4^\circ & v_4^\circ & w_4^\circ \\ \left[\begin{array}{ccc} 1.0 & 0 & 0.207 \\ 0 & 0.5 & 0 \\ 0.207 & 0 & 1.5 \end{array}\right] \end{array}\begin{array}{c} u_4^\circ \\ v_4^\circ \\ w_4^\circ \end{array} \tag{2.73}$$

Now the vector of loads corresponding to the free displacements $\{q_F\}$ is obtained by considering the external loads in the directions of each of these free displacements, as follows:-

$$\{q_F\} = \begin{Bmatrix} 3\ \text{kN} \\ -2\ \text{kN} \\ -1\ \text{kN} \end{Bmatrix} \begin{array}{c} u_4^\circ \\ v_4^\circ \\ w_4^\circ \end{array} \tag{2.74}$$

Now the vector of free displacements, namely, $\{u_F\}$ is obtained from the equation:-

$$\{q_F\} = [K_{11}]\ \{u_F\}$$

$$\text{or} \begin{Bmatrix} 3 \\ -2 \\ -1 \end{Bmatrix} = \frac{AE}{10}\begin{bmatrix} 1.0 & 0 & 0.207 \\ 0 & 0.5 & 0 \\ 0.207 & 0 & 1.5 \end{bmatrix}\begin{Bmatrix} u_4^\circ \\ v_4^\circ \\ w_4^\circ \end{Bmatrix} \tag{2.75}$$

Expanding equation (2.75) into three simultaneous equations, the following is obtained:

$$\begin{aligned} 3 &= \frac{AE}{10}\left(u_4^\circ + 0 + 0.207 w_4^\circ\right) \\ -2 &= \frac{AE}{10}\left(0 + 0.5\ v_4^\circ + 0\right) \\ -1 &= \frac{AE}{10}\left(0.207\ u_4^\circ + 0 + 1.5\ w_4^\circ\right) \end{aligned} \tag{2.76}$$

From equation (2.76 b)

$$v_4^\circ = \frac{-2 \times 10}{AE \times 0.5} = -40/AE \tag{2.77}$$

Multiplying equation (2.76a) by 0.207, the following is obtained.

$$0.621 = \frac{AE}{10} (0.207 \, u_4^\circ + 0.0429 w_4^\circ) \qquad (2.78)$$

Taking equation (2.78) from equation (2.76c),

$$-1.621 = \frac{AE}{10} \times 1.457 \, w_4^\circ$$

or $w_4^\circ = -11.126/AE$ \hfill (2.79)

Substituting equation (2.79) into equation (2.76a),

$$u_4^\circ = \frac{30}{AE} - 0.207 \times (-11.126/AE)$$

$$\underline{u_4^\circ = 32.3/AE}$$

To determine the elemental forces in each member, it will be necessary to resolve these global displacements along the local x axes of each element in turn, and calculate how much that member increases or decreases it length. Once this is known, the force in that member can be calculated from Hooke's law, as follows:

Member 1-4

From equation (2.58) the deflection of element 1-4 along its local axis, at node 4, is given by:-

$$u_4 = [C_{x,x^\circ} \quad C_{x,y^\circ} \quad C_{x,y^\circ}] \begin{Bmatrix} u_4^\circ \\ v_4^\circ \\ w_4^\circ \end{Bmatrix} \qquad (2.80)$$

Substituting equation (2.67) into (2.80),

$$u_4 = [0.5 \quad 0.5 \quad 0.707] \, \frac{1}{AE} \begin{Bmatrix} 32.3 \\ -40.0 \\ -11.126 \end{Bmatrix}$$

$$\underline{u_4 = -11.716/AE}$$

Now $u_1 = 0$, hence from Hooke's law or equation (2.4), the force in element 1-4 is given by:-

$$F_{1-4} = AE \, (u_4 - u_1)/\ell_{1-4} = AE \, (-11.716/AE)/10$$

$$F_{1-4} = -1.172 \text{ kN (compressive)}$$

Member 2-4

Now $u_4 = [C_{x,x°} \quad C_{x,y°} \quad C_{x,z°}] \begin{Bmatrix} u_4° \\ v_4° \\ w_4° \end{Bmatrix}$

and in this case, the directional cosines are given by equation (2.69).

$$\therefore u_4 = [-0.707 \quad 0 \quad 0.707] \frac{1}{AE} \begin{Bmatrix} 32.3 \\ -40 \\ -11.126 \end{Bmatrix}$$

$$u_4 = -30.702/AE$$

Now $u_2 = 0$; hence, from Hooke's law, the force in element 2-4 is given by

$F_{2-4} = AE(u_4 - u_2)/\ell_{2-4}$

$= AE(-30.702/AE - 0)/10$

$\underline{F_{2-4} = -3.07 \text{ kN (compressive)}}$

Member 4-3

Now $u_4 = [C_{x,x°} \quad C_{x,y°} \quad C_{x,z°}] \begin{Bmatrix} u_4° \\ v_4° \\ w_4° \end{Bmatrix}$

In this case, the directional cosines are given by equation (2.71).

$$u_4 = [-0.5 \quad 0.5 \quad -0.707] \frac{1}{AE} \begin{Bmatrix} 32.3 \\ -40 \\ -11.126 \end{Bmatrix}$$

$\underline{u_4 = -28.28/AE}$

Now $u_3 = 0$; hence from Hooke's law, the force in element 4-3 is given by:-

$F_{4-3} = AE(u_3 - u_4)/\ell_{4-3} = AE(0 + 28.28/AE)/10$

$\underline{F_{4-3} = 2.828 \text{ kN (tensile)}}$

2.5 Continuous Beams

The elemental stiffness matrix for a beam element can be obtained by considering the beam element of Figure 2.13.

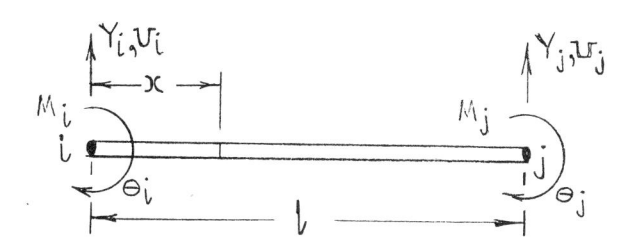

Figure 2.13 Beam element.

In the horizontal beam element of Figure 2.11, there are four degrees of freedom, namely, v_i, θ_i, v_j and θ_j; all are shown positive.

Additionally,

Y_i = vertical reaction at node i

Y_j = vertical reaction at node j

M_i = couple at node i; clockwise positive

M_j = couple at node j; clockwise positive

Now from references [3 and 4],

$$EI \frac{d^2v}{dx^2} = M = Y_i x + M_i \qquad (2.81)$$

$$\& \; EI \frac{dv}{dx} = Y_i x^2/2 + M_i x + A \qquad (2.82)$$

$$\& \; EI \, v = Y_i x^3/6 + M_i x^2/2 + Ax + B \qquad (2.83)$$

Now, there are four unknowns in equation (2.83), namely, Y_i, M_i, A and B, hence, four boundary conditions, or boundary values are required; these are

@ $x = 0$, $v = v_i$ and $\theta = \theta_i = \left(-\dfrac{dv}{dx}\right)_{x=0}$

&

@ $x = \ell$, $v = v_j$ and $\theta = \theta_j = \left(-\dfrac{dv}{dx}\right)_{x=\ell}$

From the first two boundary conditions, we get the following two simultaneous equations:-

$EI\, v_i = B$

and $EI\theta_i = -A$

or $B = EI\, v_i$ \hfill (2.84)

and $A = -EI\, \theta_i$ \hfill (2.85)

From the second two boundary conditions, we get the following two simultaneous equations:

$EI\, v_j = Y_i \ell^3/6 + M_i \ell^2/2 - EI\theta_i \ell + EI\, v_i$ \hfill (2.86)

and $EI\theta_j = -Y_i\ell^2/2 - M_i\ell + EI\, \theta_i$ \hfill (2.87)

Multiply equation (2.87) by $\ell/3$ to give $EI\theta_j\ell/3 = -Y_i\ell^3/6 - M_i\ell^2/3 + EI\theta_i\ell/3$ \hfill (2.88)

Adding equations (2.86) and (2.88) will result in eliminating Y_i, as follows:

$EI\,(v_j + \theta_j\ell/3) = M_i\ell^2\,(1/2 - 1/3) - EI\theta_i\,(\ell - \ell/3) + EI\, v_i$

or $EI\,(v_j + \theta_j\ell/3 - v_i + 2\theta_i\ell/3) = M_i\ell^2/6$

or $M_i = \dfrac{6EI}{\ell^2}\,[(v_j - v_i) + (2\theta_i + \theta_j)\,\ell/3]$ \hfill (2.89)

Substituting equation (2.89) into equation (2.87)

$Y_i\ell^2/2 = -EI\theta_j - \dfrac{6EI}{\ell}\,[(v_j - v_i) + (2\theta_i + \theta_j)\ell/3] + EI\theta_i$

or $Y_i = -\dfrac{12EI}{\ell^3}\,(v_j - v_i) + \dfrac{EI}{\ell^2}\,(-2\theta_j + 2\theta_i - 8\theta_i - 4\theta_j)$

$= \dfrac{12EI}{\ell^3}\,(v_i - v_j) - \dfrac{EI}{\ell^2}\,(6\theta_i + 6\theta_j)$

or $Y_i = \dfrac{12EI}{\ell^2}\,(v_i - v_j) - \dfrac{6EI}{\ell^2}\,(\theta_i + \theta_j)$ \hfill (2.90)

Resolving vertically, $Y_j = -Y_i$

$$Y_j = \frac{-12EI}{\ell^3}(v_i - v_j) + \frac{6EI}{\ell^2}(\theta_i + \theta_j) \qquad (2.91)$$

To determine M_j, take moments about the j node.

$M_i + M_j + Y_i\ell = 0$

or $M_j = -M_i - Y_i\ell$

$$= \frac{6EI}{\ell^2}[(v_j - v_i) + (\theta_i + 2\theta_j)\ell/3] \qquad (2.92)$$

Rewriting the slope-deflection equations (2.89) to (2.92) in matrix form, the following is obtained:-

$$\begin{Bmatrix} Y_i \\ M_i \\ Y_j \\ M_j \end{Bmatrix} = EI \begin{bmatrix} 12/\ell^3 & -6/\ell^2 & -12/\ell^3 & -6/\ell^2 \\ -6/\ell^2 & 4/\ell & 6/\ell^2 & 2/\ell \\ -12/\ell^3 & 6/\ell^2 & 12/\ell^3 & 6/\ell^2 \\ -6/\ell^2 & 2/\ell & 6/\ell^2 & 4/\ell \end{bmatrix} \begin{Bmatrix} v_i \\ \theta_i \\ v_j \\ \theta_j \end{Bmatrix} \qquad (2.93)$$

or $\{P_i\} = [k]\{U_i\}$

or $[k]$ = the elemental stiffness matrix of a beam

$$[k] = EI \begin{bmatrix} 12/\ell^3 & -6/\ell^2 & -12/\ell^3 & -6/\ell^2 \\ -6/\ell^2 & 4/\ell & 6/\ell^2 & 2/\ell \\ -12/\ell^3 & 6/\ell^2 & 12/\ell^3 & 6/\ell^2 \\ -6/\ell^2 & 2/\ell & 6/\ell^2 & 4/\ell \end{bmatrix} \begin{matrix} v_i \\ \theta_i \\ v_j \\ \theta_j \end{matrix} \qquad (2.94)$$

with column labels $v_i \quad \theta_i \quad v_j \quad \theta_j$

Example 2.4

Determine the nodal displacements and bending moments for the encastré beam of Figure 2.14.

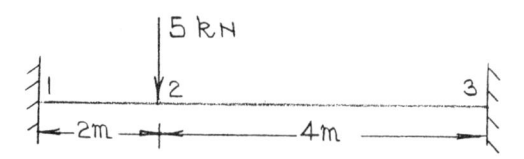

```
           5 kN
    1       ↓2                    3
   ╱|_____|_____|╲
   ╱|  2m    |         4m         |╲
         
```

Figure 2.14 - Encastré beam.

For beam elements, the elements must be numbered from left to right.

Element 1-2

$i = 1$ and $j = 2$, $\ell = 2m$

Substituting ℓ into equation (2.94), and removing the columns and rows corresponding to the zero displacements, which in this case are v_1 and θ_1, the following is obtained for the stiffness matrix for element 1-2.

$$[k_{1\text{-}2}]\ EI \quad \begin{array}{cccc} v_1 & \theta_1 & v_2 & \theta_2 \end{array} \begin{bmatrix} & & & \\ & & & \\ & & 12/2^3 & 6/2^2 \\ & & 6/2^2 & 4/2 \end{bmatrix} \begin{array}{c} v_1 \\ \theta_1 \\ v_1 \\ \theta_1 \end{array}$$

$$= EI \begin{bmatrix} 1.5 & 1.5 \\ 1.5 & 2 \end{bmatrix} \begin{array}{c} v_2 \\ \theta_2 \end{array} \qquad (2.95)$$

Element 2-3

$i = 2$, $j = 3$, $\ell = 4\ m$

Substituting ℓ into equation (2.94) and removing the columns and rows corresponding to the zero displacements, which in this case, are v_3 and θ_3, the following is obtained for the stiffness matrix of element 2-3.

$$[k_{2\text{-}3}] = EI \begin{bmatrix} v_2 & \theta_2 & v_3 & \theta_3 \\ 12/4^3 & -6/4^2 & & \\ -6/4^2 & 4/4 & & \\ \hline & & & \\ & & & \end{bmatrix} \begin{matrix} v_2 \\ \theta_2 \\ v_3 \\ \theta_3 \end{matrix}$$

$$= EI \begin{bmatrix} v_2 & \theta_2 \\ 0.1875 & -0.375 \\ -0.375 & 1 \end{bmatrix} \begin{matrix} v_2 \\ \theta_2 \end{matrix} \qquad (2.96)$$

To obtain that part of the system matrix $[K_{11}]$, that corresponds to the free displacements v_2 and θ_2, the coefficients of the two elemental stiffness matrices, corresponding to these displacements, must be added together, as follows:-

$$[K_{11}] = EI \begin{bmatrix} v_2 & \theta_2 \\ 1.5 + 0.1875 & 1.5 - 0.375 \\ 1.5 - 0.375 & 2.0 + 1.0 \end{bmatrix} \begin{matrix} v_2 \\ \theta_2 \end{matrix}$$

$$= EI \begin{bmatrix} v_2 & \theta_2 \\ 1.6875 & 1.125 \\ 1.125 & 3 \end{bmatrix} \begin{matrix} v_2 \\ \theta_2 \end{matrix} \qquad (2.97)$$

Now $[K_{11}]^{-1} = \dfrac{\dfrac{1}{EI}\begin{bmatrix} 3 & -1.125 \\ -1.125 & 1.6875 \end{bmatrix}}{1.6875 \times 3 - 1.125 \times 1.125}$

or $[K_{11}]^{-1} = \dfrac{1}{EI} \begin{bmatrix} 0.790 & -0.296 \\ -0.296 & 0.444 \end{bmatrix}$

& $\{u_F\} = \begin{Bmatrix} v_2 \\ \theta_2 \end{Bmatrix} = [K_{11}]^{-1} \{q_F\}$

where the vector of loads $\{q_F\}$ is obtained by considering the forces and couples in the directions of the free displacements v_2 and θ_2, as follows:

$\{q_F\} = \begin{Bmatrix} -5 \text{ kN} \\ 0 \text{ kN.m} \end{Bmatrix} \begin{matrix} v_2 \\ \theta_2 \end{matrix}$

Hence,

$\begin{Bmatrix} v_2 \\ \theta_2 \end{Bmatrix} = \dfrac{1}{EI} \begin{bmatrix} 0.790 & -0.296 \\ -0.296 & 0.444 \end{bmatrix} \begin{Bmatrix} -5 \\ 0 \end{Bmatrix}$

$\begin{Bmatrix} v_2 \\ \theta_2 \end{Bmatrix} = \dfrac{1}{EI} \begin{Bmatrix} -3.95 \\ 1.48 \end{Bmatrix}$ \hfill (2.98)

To determine the nodal moments, the slope-deflection equations, namely equations (2.89) and (2.92), have to be used as follows:-

Element 1-2

$i = 1$, $j = 2$, $v_1 = \theta_1 = 0$ and $v_2 = -3.95/EI$ and $\theta_2 = 1.48/EI$

Substituting the above into equation (2.89)

$$M_1 = \frac{6EI}{4} [(-3.95/EI - 0) + (0 + 1.48/EI)2/3]$$

$\underline{M_1 = -4.445 \text{ kN.m}}$

Substituting the displacements into equation (2.92)

$$M_2 = \frac{6EI}{4} [(-3.95/EI - 0) + (0 + 2 \times 1.48/EI) 2/3]$$

$\underline{M_2 = -2.965 \text{ kN.m}}$

Element 2-3

$i = 2, j = 3 \quad v_2 = -3.95/EI, \quad \theta_2 = 1.48/EI, \quad v_3 = \theta_3 = 0$

Substituting the above displacements into the slope-deflection equation (2.89)

$$M_2 = \frac{6EI}{16} [(0 + 3.95/EI) + (2 \times 1.48/EI + 0)4/3]$$

$\underline{M_2 = 2.961 \text{ kNm}}$

Substituting the above displacements into the slope-deflection equation (2.92),

$$M_3 = \frac{6EI}{16} [(0 + 3.95/EI) + (1.48/EI + 0) 4/3]$$

$\underline{M_3 = 2.22 \text{ kN.m}}$

NB The slight differences between the magnitudes for the two values for M_2 were due to round-off error.

Example 2.5

Determine the nodal displacements and bending moments for the continuous beams of Figure 2.15, which is subjected to a downward uniformly distributed load.

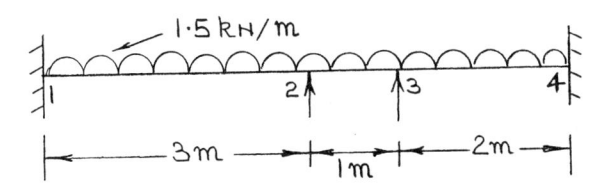

Figure 2.15 Continuous beam.

The main difficulty of analysing the continuous beam of Figure 2.15 is in obtaining the equivalent nodal forces for the distributed load; this is necessary as the finite element method can only cope with nodal forces. To deal with distributed loads on beams and frames, the following process, which is based on superposition, is used.

(a) Fix the beam or frame at all the nodes of joints, and calculate the end fixing "forces" to achieve this.

(b) The beam or frame in condition (a) is not in equilibrium, hence, to keep it in equilibrium, subject the beam or frame to the negative resultants of the end fixing "forces".

(c) Use the negative resultants of the end fixing "forces", as the external load vector, and calculate the nodal displacements and bending moments due to this vector.

(d) Now as the beam or frame was fixed at its nodes in condition (a), it will be necessary to superimpose the end fixing moments with the nodal moments calculated in condition (c) to give the final values of the nodal moments.

To calculate $\{q_F\}$, the beam will be fixed at all nodes from 1 to 4, as follows:-

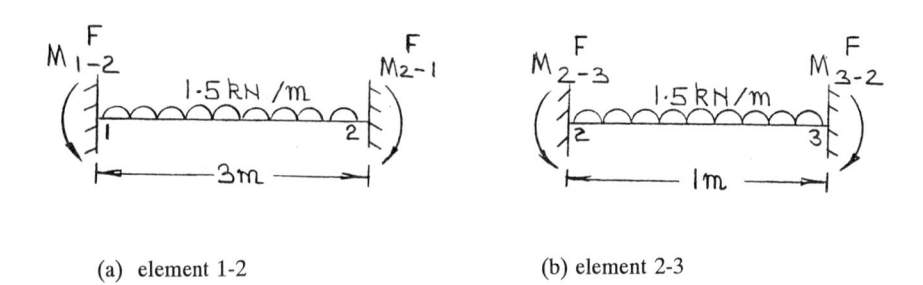

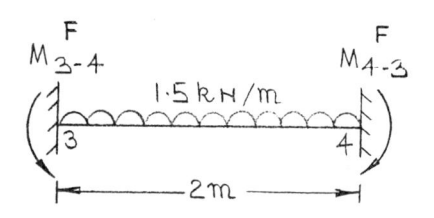

(a) element 1-2

(b) element 2-3

(c) element 3-4

Figure 2.16 Beam elements with end fixing moments.

Sec. 2.5] Continuous Beams 63

$$M^F_{1-2} = \frac{-w\ell^2}{12} = \frac{-1.5 \times 3^2}{12} = -1.125 \text{ kN.m}$$

$$M^F_{2-1} = \frac{w\ell^2}{12} = \frac{1.5 \times 3^2}{12} = 1.125 \text{ kN.m}$$

$$M^F_{2-3} = \frac{-w\ell^2}{12} = \frac{-1.5 \times 1^2}{12} = -0.125 \text{ kN.m}$$

(2.99)

$$M^F_{3-2} = \frac{w\ell^2}{12} = \frac{1.5 \times 1^2}{12} = 0.125 \text{ kN.m}$$

$$M^F_{3-4} = \frac{-w\ell^2}{12} = \frac{-1.5 \times 2^2}{12} = -0.5 \text{ kN.m}$$

$$M^F_{4-3} = \frac{w\ell^2}{12} = 0.5 \text{ kN.m}$$

Now $v_1 = \theta_1 = v_2 = v_3 = v_4 = \theta_4 = 0$, hence, the only unbalanced moments that need to be considered are in the directions θ_2 and θ_3 as shown in Figure 2.17(a); the negative resultants, corresponding to the free displacements θ_2 and θ_3, are shown in Figure 2.17(b)

1.125 − 0.125
= 1 kN m

−0.5 + 0.125
= −0.375 kN m

(a) Resultants in directions of θ_2 and θ_3

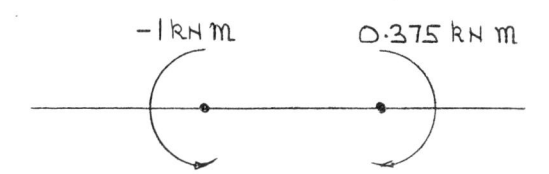

−1 kN m   0.375 kN m

(b) Negative resultants corresponding to θ_2 and θ_3

Figure 2.17 Resultants and negative resultants.

From Figure 2.17 (b)

$$\{q_F\} = \begin{Bmatrix} -1 \\ 0.375 \end{Bmatrix} \begin{matrix} \theta_2 \\ \theta_3 \end{matrix} \tag{2.100}$$

Element 1-2

$i = 1, j = 2, \ell = 3m$

Substituting ℓ into equation (2.94), and removing the rows and columns corresponding to the zero displacements, which in this case are v_1, θ_1 and v_2,

$$[k_{1\text{-}2}] = EI \begin{bmatrix} & v_1 & \theta_1 & v_2 & \theta_2 \\ \hline & & & & \\ \hline & & & & \\ \hline & & & 4/3 & \\ \end{bmatrix} \begin{matrix} v_1 \\ \theta_1 \\ v_2 \\ \theta_2 \end{matrix} = EI\,[1.333]\,\theta_2 \tag{2.101}$$

Element 2-3

$i = 2, j = 3, \ell = 1m$

Substituting ℓ into equation (2.94), and removing the rows and columns corresponding to the zero displacements, which in this case are v_2 and v_3

$$[k_{2\text{-}3}] = EI \begin{bmatrix} & v_2 & \theta_2 & v_3 & \theta_3 \\ \hline & & 4/1 & & 2/1 \\ \hline & & 2/1 & & 4/1 \\ \end{bmatrix} \begin{matrix} v_2 \\ \theta_2 \\ v_3 \\ \theta_3 \end{matrix} \tag{2.102}$$

Element 3-4

$i = 3, j = 4, \ell = 2m$

Substituting ℓ into equation (2.94), and removing the rows and columns corresponding to the zero displacements, which in this case, are v_3, v_4 and θ_4

$$[k_{3\text{-}4}] = EI \begin{bmatrix} & v_3 & \theta_3 & v_4 & \theta_4 & \\ & & & & & \\ & & 4/2 & & & \\ & & & & & \\ & & & & & \\ & & & & & \end{bmatrix} \begin{matrix} v_3 \\ \theta_3 \\ v_4 \\ \theta_4 \end{matrix} \qquad (2.103)$$

To obtain that part of the structural stiffness matrix, $[K_{11}]$, that corresponds to the free displacements, namely θ_2 and θ_3, the appropriate coefficients of the elemental stiffness matrices of equations (2.101) to (2.103) must be added together, as follows:

$$[K_{11}] = EI \begin{bmatrix} \theta_2 & \theta_3 \\ 1.333 + 4 & 2 \\ 2 & 4 + 2 \end{bmatrix} \begin{matrix} \theta_2 \\ \theta_3 \end{matrix} \qquad (2.104)$$

$$[K_{11}] = EI \begin{bmatrix} \theta_2 & \theta_3 \\ 5.333 & 2 \\ 2 & 6 \end{bmatrix} \begin{matrix} \theta_2 \\ \theta_3 \end{matrix} \qquad (2.105)$$

$$\text{Now } [K_{11}]^{-1} = \frac{\dfrac{1}{EI} \begin{bmatrix} 6 & -2 \\ -2 & 5.333 \end{bmatrix}}{5.333 \times 6 - 2 \times 2}$$

$$= \frac{1}{EI} \begin{bmatrix} 0.214 & -0.0714 \\ -0.0714 & 0.190 \end{bmatrix} \qquad (2.106)$$

$$\text{Now } \{u_F\} = \begin{Bmatrix} \theta_2 \\ \theta_3 \end{Bmatrix} = [K_{11}]^{-1} \{q_F\} \qquad (2.107)$$

Hence, from equations, (2.100) and (2.106)

$$\begin{Bmatrix} \theta_2 \\ \theta_3 \end{Bmatrix} = \frac{1}{EI} \begin{bmatrix} 0.214 & -0.0714 \\ -0.0714 & 0.190 \end{bmatrix} \begin{Bmatrix} -1.0 \\ 0.375 \end{Bmatrix}$$

$$= \frac{1}{EI} \begin{Bmatrix} -0.241 \\ 0.143 \end{Bmatrix} \tag{2.108}$$

To determine the nodal moments, the nodal displacements will be substituted into equations (2.89) and (2.92), and then the end fixing moments will be added, as follows:

Element 1-2

$i = 1, j = 2, v_1 = \theta_1 = v_2 = 0; \theta_2 = -0.241/EI$

Substituting the above data into the slope deflection equation (2.89).

$$M_1 = \frac{6EI}{9} [(0 - 0) + (0 - 0.241/EI)3/3] - 1.125$$

$\underline{M_1 = -1.286 \text{ kN.m}}$

Similarly, from equation (2.92)

$$M_2 = \frac{6EI}{9} [(0 - 0) + (0 - 2 \times 0.241/EI)3/3] + 1.125$$

$\underline{M_2 = 0.804 \text{ kN.m}}$

Element 2-3

$i = 2, j = 3, v_2 = v_3 = 0; \theta_2 = -0.241/EI$ and $\theta_3 = 0.143/EI$

Substituting the above data into equation (2.89)

$$M_2 = \frac{6EI}{1} [(0 - 0) + (-2 \times 0.241/EI + 0.143/EI)1/3] - 0.125$$

$\underline{M_2 = -0.803 \text{ kN.m}}$

Similarly, from equation (2.92)

$$M_3 = \frac{6EI}{1} [(0 - 0) + (-0.241/EI + 2 \times 0.143/EI) 1/3] + 0.125$$

$\underline{M_3 = 0.215 \text{ kN.m}}$

Element 3-4

$i = 3, j = 4, v_3 = v_4 = \theta_4 = 0, \theta_3 = 0.143/EI$

Substituting the above data into equation (2.89)

$$M_3 = \frac{6EI}{4} [(0 - 0) + (2 \times 0.143/EI + 0)2/3] - 0.5$$

$\underline{M_3 = -0.214 \text{ kN.m}}$

Similarly, from equation (2.92)

$$M_4 = \frac{6EI}{4} [(0 - 0) + (0.143/EI + 0)]2/3] + 0.5$$

$\underline{M_4 = 0.643 \text{ kN.m}}$

2.6 Rigid-Jointed Plane Frames

The element for analysing rigid-jointed plane frames is a combination of the plane pin-jointed truss element and the inclined beam element, as shown in Figure 2.18.

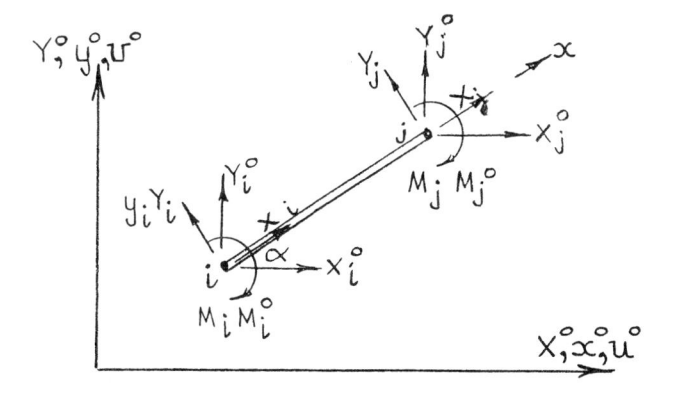

Figure 2.18 Plane frame element.

From equations (2.28) and (2.93) the "force-displacement" relationships for a frame element, in local co-ordinates, are given by equation (2.109).

From equations (2.28) and (2.93) the "force-displacement" relationships for a frame element, in local co-ordinates, are given by equation (2.109)

$$\begin{Bmatrix} X_i \\ Y_i \\ M_i \\ X_j \\ Y_j \\ M_j \end{Bmatrix} = EI \begin{bmatrix} (A/\ell I) & 0 & 0 & (-A/\ell I) & 0 & 0 \\ 0 & 12/\ell^3 & -6/\ell^2 & 0 & -12/\ell^3 & -6/\ell^2 \\ 0 & -6/\ell^2 & 4/\ell & 0 & 6/\ell^2 & 2/\ell \\ (-A/\ell I) & 0 & 0 & (A/\ell I) & 0 & 0 \\ 0 & -12/\ell^3 & 6/\ell^2 & 0 & 12/\ell^3 & 6/\ell^2 \\ 0 & -6/\ell^2 & 2/\ell & 0 & 6/\ell^2 & 4/\ell \end{bmatrix} \begin{Bmatrix} u_i \\ v_i \\ \theta_i \\ u_j \\ v_j \\ \theta_j \end{Bmatrix} \quad (2.109)$$

or $\{P_i\} = [k]\{U_i\}$

Now the relationship between local and global axes, is given by:

$$\begin{Bmatrix} X_i \\ Y_i \\ M_i \\ X_j \\ Y_j \\ M_j \end{Bmatrix} = \left[\begin{array}{c|c} \begin{matrix} c & s & 0 \\ -s & c & 0 \\ 0 & 0 & 1 \end{matrix} & O_3 \\ \hline O_3 & \begin{matrix} c & s & 0 \\ -s & c & 0 \\ 0 & 0 & 1 \end{matrix} \end{array} \right] \begin{Bmatrix} X_i^\circ \\ Y_i^\circ \\ M_i^\circ \\ X_j^\circ \\ Y_j^\circ \\ M_j^\circ \end{Bmatrix} \quad (2.110)$$

or

$$\{P_i\} = \left[\begin{array}{c|c} T & O_3 \\ \hline O_3 & T \end{array} \right] \{P_i^\circ\} \quad (2.111)$$

Now, from equation (2.43),

$$[k^\circ] = \begin{bmatrix} T & O_3 \\ O_3 & T \end{bmatrix}^T [k] \begin{bmatrix} T & O_3 \\ O_3 & T \end{bmatrix}$$

$$= [k_b^\circ] + [k_r^\circ] \quad (2.112)$$

Sec. 2.6] Rigid-Jointed Plane Frames

where

$$[k_b^\circ] = EI \begin{bmatrix} 12s^2/\ell^3 & & & & \text{symmetrical} & \\ -12cs/\ell^3 & 12c^2/\ell^3 & & & & \\ 6s/\ell^2 & -6c/\ell^2 & 4/\ell & & & \\ -12s^2/\ell^3 & 12cs/\ell^3 & -6s/\ell^2 & 12s^2/\ell^3 & & \\ 12cs/\ell^3 & -12c^2/\ell^3 & 6c/\ell^2 & -12cs/\ell^3 & 12c^2/\ell^3 & \\ 6s/\ell^2 & -6c/\ell^2 & 2/\ell & -6s/\ell^2 & 6c/\ell^2 & 4/\ell \end{bmatrix} \begin{matrix} u_i^\circ \\ v_i^\circ \\ \theta_i \\ u_j^\circ \\ v_j^\circ \\ \theta_j \end{matrix}$$

(column labels: $u_i^\circ \; v_i^\circ \; \theta_i \; u_j^\circ \; v_j^\circ \; \theta_j$)

(2.113)

= stiffness matrix for an inclined beam element.

$$[k_r^\circ] = \frac{AE}{\ell} \begin{bmatrix} c^2 & cs & 0 & -c^2 & -cs & 0 \\ cs & s^2 & 0 & -cs & -s^2 & 0 \\ 0 & 0 & 0 & 0 & 0 & 0 \\ -c_2 & -cs & 0 & c^2 & cs & 0 \\ -cs & -s^2 & 0 & cs & s^2 & 0 \\ 0 & 0 & 0 & 0 & 0 & 0 \end{bmatrix} \begin{matrix} u_i^\circ \\ v_i^\circ \\ \theta_i \\ u_j^\circ \\ v_j^\circ \\ \theta_j \end{matrix}$$

(column labels: $u_i^\circ \; v_i^\circ \; \theta_i \; u_j^\circ \; v_j^\circ \; \theta_j$)

(2.114)

= stiffness matrix for an inclined rod element

$c = \cos \alpha$
$s = \sin \alpha$

Example 2.6

Determine the nodal displacements and bending moments for the rigid-jointed plane frame of Figure 2.19. It may be assumed that the axial stiffness is very large in comparison with the flexural stiffness, so that,

$u_2^\circ = u_3^\circ$ and $v_2^\circ = v_3^\circ = 0$

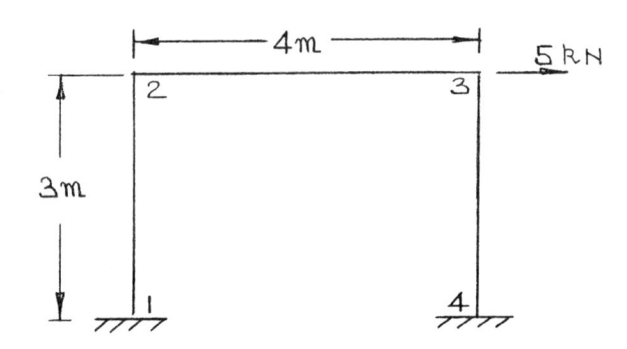

Figure 2.19 Rigid-jointed plane frame.

As $u_2^\circ = u_3^\circ$ and $v_2 = v_3^\circ$, it will be quite reasonable to neglect the influence of equation (2.114).

Element 1-2

$i = 1, j = 2 \ \alpha = 90°, \ \ell_{1-2} = 3\text{m}, \ c = 0, \ s = 1$

Substituting the above data into equation (2.113) and removing the rows and columns corresponding to the zero displacements, which in this case are u_1°, v_1°, θ_1, and v_2°, the elemental stiffness matrix for element 1-2, is

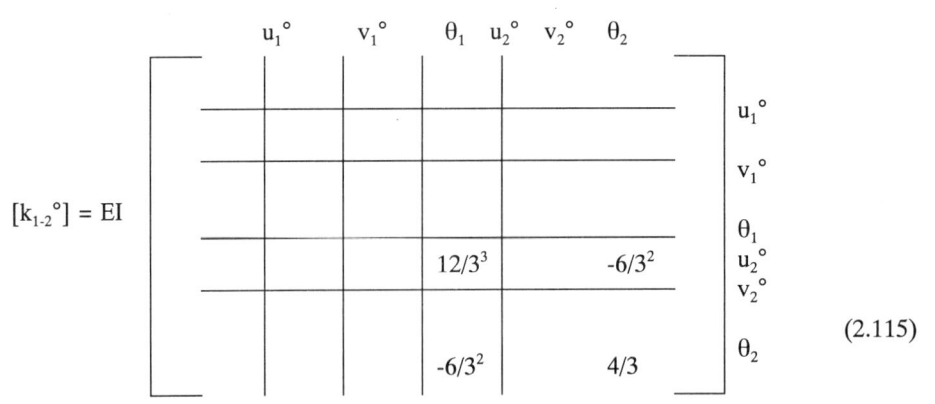

$$[k_{1-2}^\circ] = EI \begin{bmatrix} & & & & & \\ & & & & & \\ & & & & & \\ & & 12/3^3 & & -6/3^2 & \\ & & & & & \\ & & -6/3^2 & & 4/3 & \end{bmatrix} \begin{matrix} u_1^\circ \\ v_1^\circ \\ \theta_1 \\ u_2^\circ \\ v_2^\circ \\ \theta_2 \end{matrix} \quad (2.115)$$

with column headers $u_1^\circ \ v_1^\circ \ \theta_1 \ u_2^\circ \ v_2^\circ \ \theta_2$

Sec. 2.6] Rigid-Jointed Plane Frames 71

$$
\text{or } [k_{1\text{-}2}°] = EI \begin{bmatrix} u_2° & \theta_2 \\ 0.444 & -0.667 \\ -0.667 & 1.333 \end{bmatrix} \begin{matrix} u_2° \\ \theta_2 \end{matrix} \quad (2.116)
$$

Element 2-3

$i = 2, j = 3, \alpha = 0°, \ell_{2\text{-}3} = 4m; c = 1; s = 0$

Substituting the above data into equation (2.113) and removing the columns and rows corresponding to the zero displacements, which in this case are v_2 and v_3, the stiffness matrix for element 2-3 is given by:

$$
= \begin{bmatrix} u_2° & v_2° & \theta_2° & u_3° & v_3° & \theta_3 \\ 0 & & 0 & 0 & & 0 \\ 0 & & 4/4 & 0 & & 2/4 \\ 0 & & 0 & 0 & & 0 \\ 0 & & 2/4 & 0 & & 4/4 \end{bmatrix} \begin{matrix} u_2° \\ v_2° \\ \theta_2° \\ u_2 \\ v_3° \\ \theta_3° \end{matrix} \quad (2.117)
$$

$$
= EI \begin{bmatrix} \theta_2 & \theta_3 \\ 1 & 0.5 \\ 0.5 & 1 \end{bmatrix} \begin{matrix} \theta_2 \\ \theta_3 \end{matrix} \quad (2.118)
$$

Element 3-4

$i = 3, j = 4, \alpha = -90°, \ell_{3\text{-}4} = 3m, c = 0, s = -1$

Substituting the above data into equation (2.113) and removing the rows and columns corresponding to the zero displacements, which in this case are $v_3°$, $u_4°$, $v_4°$ and θ_4, the stiffness matrix for element 3-4 is given by:

$$[k_{3-4}{}^o] = EI \begin{bmatrix} & u_3{}^o & v_3{}^o & \theta_3{}^o & u_4{}^o & v_4{}^o & \theta_4 & \\ & 12/3^3 & -6/3^2 & & & & & \\ & -6/3^2 & 4/3 & & & & & \\ & & & & & & & \\ & & & & & & & \\ & & & & & & & \\ & & & & & & & \end{bmatrix} \begin{matrix} u_3{}^o \\ v_3{}^o \\ \theta_3{}^o \\ u_4 \\ v_4{}^o \\ \theta_4{}^o \end{matrix} \quad (2.119)$$

$$= EI \begin{bmatrix} u_3{}^o & \theta_3 \\ 0.444 & -0.667 \\ -0.667 & 1.333 \end{bmatrix} \begin{matrix} u_3{}^o \\ \theta_3 \end{matrix} \quad (2.120)$$

To obtain that part of the stiffness matrix $[K_{11}]$, that corresponds to the free displacements, namely, $u_2{}^o$, θ_2, $u_3{}^o$ and θ_3, the appropriate parts of the elemental stiffness matrices of equations (2.116), (2.118) and (2.120), must be added together, as shown below.

$$[K_{11}] = EI \begin{bmatrix} u_2{}^o & \theta_2 & u_3{}^o & \theta_3 \\ 0.444 & -0.667 & & \\ -0.667 & 1.333 + 1 & & 0.5 \\ & & 0.444 & -0.667 \\ & 0.5 & -0.667 & 1 + 1.333 \end{bmatrix} \begin{matrix} u_2{}^o \\ \theta_2 \\ u_3{}^o \\ \theta_3 \end{matrix}$$

$$[K_{11}] = EI \begin{bmatrix} u_2{}^o & \theta_2 & u_3{}^o & \theta_3 \\ 0.444 & -0.667 & 0 & 0 \\ -0.667 & 2.333 & 0 & 0.5 \\ 0 & 0 & 0.4444 & -0.667 \\ 0 & 0.5 & -0.667 & 2.333 \end{bmatrix} \begin{matrix} u_2{}^o \\ \theta_2 \\ u_3{}^o \\ \theta_3 \end{matrix} \quad (2.121)$$

Sec. 2.6] **Rigid-Jointed Plane Frames** 73

To obtain the vector of external loads corresponding to the free displacements, each load in the "direction" of each free displacement must be placed into the vector, as follows:

$$\{q_F\} = \begin{Bmatrix} 0 \\ 0 \\ 5 \\ 0 \end{Bmatrix} \begin{matrix} u_2^{\circ} \\ \theta_2 \\ u_3^{\circ} \\ \theta_3 \end{matrix} \tag{2.122}$$

However, the 5 kN load must be shared between nodes 2 and 3, as the elemental stiffness matrix of equation (2.114) was neglected, and as $u_2^{\circ} = u_3^{\circ}$.

$$\text{Hence } \{q_F\} = \begin{Bmatrix} 2.5 \\ 0 \\ 2.5 \\ 0 \end{Bmatrix}$$

Now $\{q_F\} = [K_{11}]\{u_F\}$

$$\therefore \begin{Bmatrix} 2.5 \\ 0 \\ 2.5 \\ 0 \end{Bmatrix} = EI \begin{bmatrix} 0.444 & -0.667 & 0 & 0 \\ -0.667 & 2.333 & 0 & 0.5 \\ 0 & 0 & 0.444 & -0.667 \\ 0 & 0.5 & -0.667 & 2.333 \end{bmatrix} \begin{Bmatrix} u_2^{\circ} \\ \theta_2 \\ u_3^{\circ} \\ \theta_3 \end{Bmatrix} \tag{2.123}$$

Expanding equation (2.123), the following four simultaneous equations are obtained:

$2.5 = EI\,(0.4442 u_2 - 0.667\,\theta_2)$

$0 = EI\,(-0.667 u_2 + 2.333\,\theta_2 + 0.5\,\theta_3)$

$2.5 = EI\,(0.444\,u_3^{\circ} - 0.667\,\theta_3)$ \hfill (2.124)

$0 = EI\,(0.5\theta_2 - 0.667\,u_3^{\circ} + 2.333\,\theta_2)$

By inspection, it can be seen that

$u_2° = u_3°$ and $\theta_2 = \theta_3$, hence, equations (2.124) can be reduced to the following two simultaneous equations:-

$$2.5 = EI (0.444 u_2° - 0.667 \theta_2) \qquad (2.125)$$

$$0 = EI (-0.667 u_2 + 2.833 \theta_2) \qquad (2.126)$$

Multiplying equation (2.125) by 0.444/0.667, the following is obtained:

$$0 = EI (-0.444 u_2° + 1.887 \theta_2) \qquad (2.127)$$

Adding equations (2.126) to (2.127), results in the following:-

$$2.5 = EI (1.22) \theta_2$$

$$\therefore \theta_2 = 2.05/EI \qquad (2.128)$$

Substituting equation (2.128) into equation 2.125,

$$u_2° = (2.5 + 1.366)/(0.444\ EI)$$

$$\underline{u_2° = 8.70/EI} \qquad (2.129)$$

Also

$\theta_3 = \theta_2 = 2.05/EI$

and

$u_3° = u_2° = 8.70/EI$

To obtain the nodal bending moments, equation (2.113) must be used, as follows:-

Element 1-2

$i = 1, j = 2, s = 1, c = 0$

From the third row of equation (2.113)

$$M_i = EI [6s/\ell^2 (u_i°) - 6c/\ell^2 (v_i°) + 4\theta_i/\ell \\ - 6s/\ell^2 (u_j°) + 6c/\ell^2 (v_j°) + 2\theta_j/\ell] \qquad (2.130)$$

For this case, $i = 1$ and $j = 2$, so that

$$M_1 = EI \left[0 - 0 + 0 - \frac{6 \times 8.70}{9\ EI} + 0 + \frac{2 \times 2.05}{3EI} \right]$$

$\underline{M_1 = -4.43\ kN.m}$

Sec. 2.6]   **Rigid-Jointed Plane Frames**   75

From the sixth row of equation (2.113)

$$M_j = EI\,[6s/\ell^2\,(u_i°) - 6c/\ell^2\,(v_i°) + 2\theta_i/\ell \\ - 6s/\ell^2\,(u_j°) + 6c/\ell^2\,(v_j°) + 4\theta_j/\ell]$$ (2.131)

or

$$M_2 = EI\left[0 - 0 - 0 - \frac{6 \times 8.70}{9EI} + 0 + \frac{4 \times 2.05}{3EI}\right]$$

$\underline{M_2 = -3.07\ \text{kN.m}}$

Element 2-3

For this case $i = 2$ and $j = 3$, $s = 0$, $c = 1$

From equation (2.130)

$$M_2 = EI\left[0 - 0 + \frac{4 \times 2.05}{4EI} - 0 + 0 + \frac{2 \times 2.05}{4EI}\right]$$

$\underline{M_2 = 3.075\ \text{kN.m}}$

From equation (2.131)

$$M_3 = EI\left[0 - 0 + \frac{2 \times 2.05}{4EI} - 0 + 0 + \frac{4 \times 2.05}{EI}\right]$$

$\underline{M_3 = 3.075\ \text{kN.m}}$

Element 3-4

For this case $i = 3$ and $j = 4$, $s = -1$, $c = 0$

From equation (2.130)

$$M_3 = EI\left[\frac{-6 \times 8.70}{9EI} + 0 + \frac{4 \times 2.05}{3EI} - 0 + 0 + 0\right]$$

$\underline{M_3 = -3.07\ \text{kN.m}}$

From equation (2.131)

$$M_4 = EI\left[\frac{-6 \times 8.70}{9EI} - 0 + \frac{2 \times 2.05}{3EI} - 0 + 0 + 0\right]$$

$\underline{M_4 = -4.43\ \text{kN.m}}$

Example 2.7

Determine the nodal displacements and bending moments for the rigid-jointed plane frame of Figure 2.20, where the axial stiffness of each element is very large compared with its flexural stiffness.

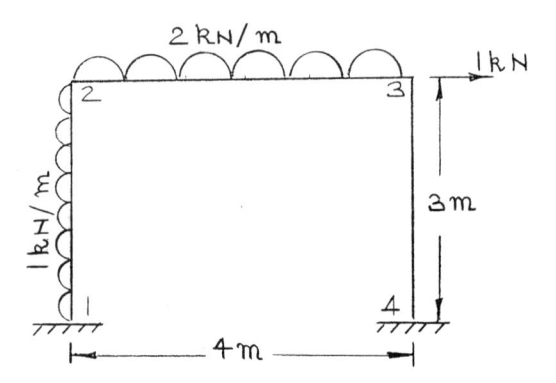

Figure 2.20 Rigid-jointed plane frame with distributed loads.

To calculate $\{q_F\}$ the vector of loads corresponding to the free displacements, namely $u_2°$, θ_2, $u_3°$ and θ_3, the frame will be firmly fixed at its nodes, and the end fixing "forces" of each element will be calculated with the aid of Figure 2.21, as follows:-

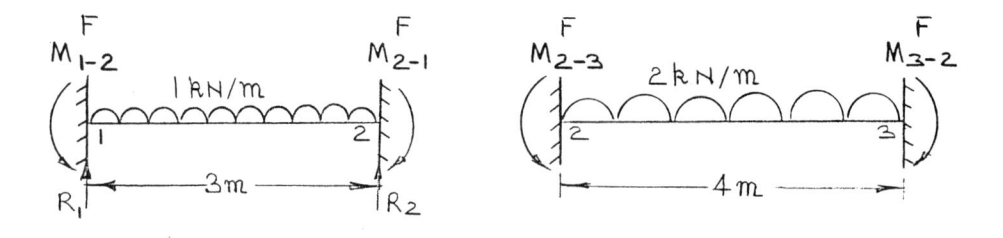

(a) Element 1-2   (b)   Element 2-3

Figure 2.21 Elements with end fixing "forces".

$$R_1 = R_2 = \frac{w\ell}{2} = \frac{1 \times 3}{2} = 1.5 \text{ kN}$$

$$M^F_{1-2} = -\frac{w\ell^2}{12} = \frac{-1 \times 3^2}{12} = -0.75 \text{ kN.m}$$

$$M^F_{2-1} = \frac{w\ell^2}{12} = 0.75 \text{ kN.m} \qquad (2.132)$$

$$M^F_{2-3} = \frac{-w\ell^2}{12} = \frac{-2 \times 4^2}{12} = -2.667 \text{ kN.m}$$

$$M^F_{3-2} = \frac{w\ell^2}{12} = \frac{2 \times 4^2}{12} = 2.667 \text{ kN.m}$$

The end fixing "forces", corresponding to the free displacements are shown in Figure 2.22.

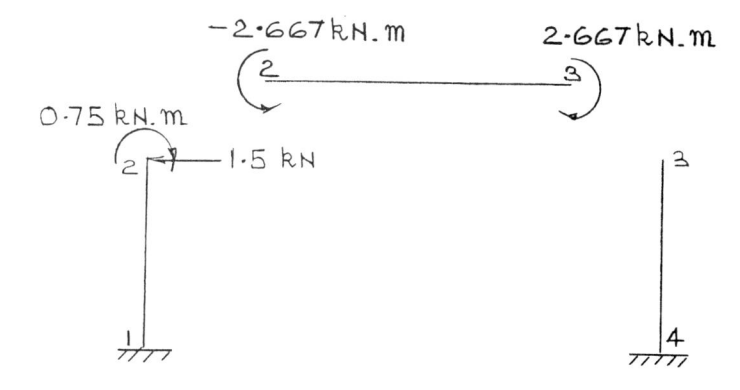

Figure 2.22 End fixing "forces".

The resultants of the end fixing "forces" are shown in Figure 2.23.

The negative resultants of the end fixing "forces", together with the 1 kN load acting at node 3, are shown in Figure 2.24.

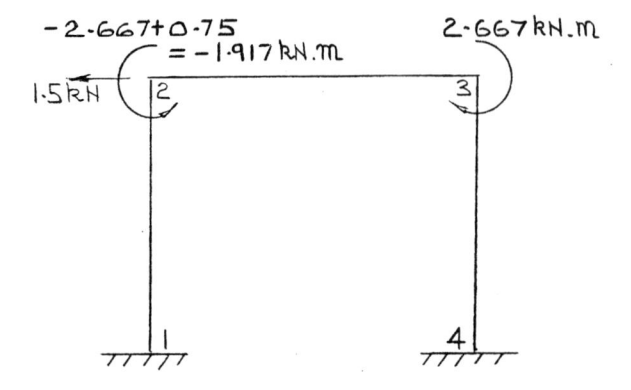

Figure 2.23 Resultants of end fixing "forces".

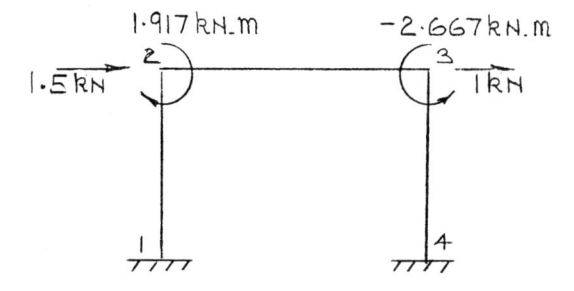

Figure 2.24 Negative resultants + 1 kN load.

Sec. 2.6] Rigid-Jointed Plane Frames

From Figure 2.24, the vector of loads $\{q_F\}$ corresponding to the free displacements, namely u_2^o, θ_2, u_3^o and θ_3 is as follows:

$$\{q_F\} = \begin{Bmatrix} 1.5 \\ 1.917 \\ 1.0 \\ -2.667 \end{Bmatrix} \begin{matrix} u_2^o \\ \theta_2 \\ u_3^o \\ \theta_3 \end{matrix} \tag{2.133}$$

From equation (2.121)

$$[K_{11}] = EI \begin{bmatrix} 0.444 & -0.667 & 0 & 0 \\ -0.667 & 2.333 & 0 & 0.5 \\ 0 & 0 & 0.444 & -0.667 \\ 0 & 0.5 & -0.667 & 2.333 \end{bmatrix} \begin{matrix} u_2^o \\ \theta_2 \\ u_3^o \\ \theta_3 \end{matrix} \tag{2.134}$$

Now,

$$\{q_F\} = [K_{11}]\{u_F\}$$

or

$$\begin{Bmatrix} 1.5 \\ 1.917 \\ 1.0 \\ -2.667 \end{Bmatrix} = EI \begin{bmatrix} 0.444 & -0.667 & 0 & 0 \\ -0.667 & 2.333 & 0 & 0.5 \\ 0 & 0 & 0.444 & -0.667 \\ 0 & 0.5 & -0.667 & 2.333 \end{bmatrix} \begin{Bmatrix} u_2^o \\ \theta_2 \\ u_3^o \\ \theta_3 \end{Bmatrix} \tag{2.135}$$

Expanding equation (2.135) results in the following four simultaneous equations:-

$1.5 = EI\ (0.444\ u_2^o - 0.667\ \theta_2)$

$1.917 = EI\ (-0.667\ u_2^o + 2.333\ \theta_2 + 0.5\ \theta_3)$ \hfill (2.136)

$1.0 = EI\ (0.444\ u_3^o - 0.667\ \theta_3)$

$-2.667 = EI\ (0.5\ \theta_2 - 0.667\ u_3^o + 2.333\ \theta_3)$

If equations (2.136) are solved in the form presented, errors will occurs, because the stiffness matrix of equation (2.114) was not considered, and also because $u_2° = u_3°$.

As $u_2° = u_3°$, it will be necessary to add together equations (2.136a) and (2.136c), as follows:

$2.5 = EI\ (0.888\ u_2° - 0.667\ \theta_2 - 0.667\ \theta_3)$

Thus, equations (2.136) are reduced to the three simultaneous equations below:

$2.5 = EI\ (0.888\ u_2° - 0.667\ \theta_2 - 0.667\ \theta_3)$

$1.917 = EI\ (-0.667\ u_2° + 2.333\ \theta_2 + 0.5\ \theta_3)$ \hfill (2.137)

$2.667 = EI\ (-0.667\ u_2° + 0.5\ \theta_2 + 2.333\ \theta_3)$

Solution of the above three simultaneous equations results in the following:

$$\begin{Bmatrix} u_2° \\ \theta_2 \\ \theta_3 \end{Bmatrix} = \frac{1}{EI} \begin{Bmatrix} 4.048 \\ 2.071 \\ -0.430 \end{Bmatrix} \quad (2.138)$$

$u_3° = u_2° = 4.048/EI$

To calculate the nodal moments, a similar process as that used for the previous problem is adopted, but the end fixing moments have to be added, as follows:

Element 1-2

$i = 1,\ j = 2,\ u_1° = v_1° = \theta_1 = v_2° = 0;\ u_2° = 4.048/EI$ and $\theta_2 = 2.071/EI,\ s = 1,\ c = 0$

Hence, from equation (2.130)

$$M_1 = EI \left(0 - 0 + 0 - \frac{6 \times 4.048}{9EI} + 0 + 2 \times \frac{2.071}{3EI} \right) - 0.75$$

$M_1 = -2.068$ kN.m

Similarly, from equation (2.131)

$$M_2 = EI \left(0 - 0 + 0 - \frac{6 \times 4.048}{9EI} + 0 + \frac{4 \times 2.071}{3EI} \right) + 0.75$$

$M_2 = 0.813$ kN.m

Element 2-3

$i = 2, j = 3, s = 0, c = 1$

Hence, from equation (2.130)

$$M_2 = EI \left[0 - 0 + \frac{4 \times 2.071}{4EI} - 0 + 0 + \frac{2 \times (-0.430)}{4EI} \right] - 2.667$$

$$\underline{M_2 = -0.881 \text{ kN.m}}$$

Similarly, from equation (2.131)

$$M_3 = EI \left(0 - 0 + \frac{2 \times 2.071}{4EI} - 0 + 0 - \frac{4 \times 0.430}{4EI} \right) + 2.667$$

$$\underline{M_3 = 3.273 \text{ kN.m}}$$

Element 3-4

$i = 3$ and $j = 4, s = -1, c = 0$

Hence, from equation (2.130)

$$M_3 = EI \left(\frac{-6 \times 4.048}{9EI} - 0 - \frac{4 \times 0.430}{3EI} - 0 + 0 + 0 \right)$$

$$\underline{M_3 = -3.272 \text{ kN.m}}$$

Similarly, from equation (2.131)

$$M_4 = EI \left(\frac{-6 \times 4.048}{9EI} - 0 - \frac{2 \times 0.430}{3EI} \right)$$

$$\underline{M_4 = -2.985 \text{ kN.m}}$$

2.7 Stiffness Matrix for a Torque Bar

The well known formula for the torsion of a uniform section bar is

$$T = GJ \frac{d\phi}{dx} \qquad (2.139)$$

where T = torque

G = rigidity

J = torsional constant

$\frac{d\phi}{dx}$ = angle of twist per unit length

Applying equation (2.139) to the torque bar of Figure 2.25, the following is obtained:

$$T_i = GJ(\phi_i - \phi_j)$$

and $\qquad (2.140)$

$$T_j = GJ(\phi_j - \phi_i)$$

where ϕ_i and ϕ_j = angles of twist at nodes i and j, respectively.

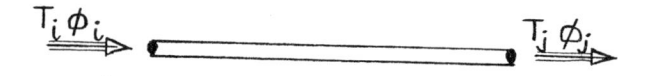

Figure 2.25 Torque bar.

Rewriting equations in matrix form, the following is obtained:

$$\begin{Bmatrix} T_i \\ T_j \end{Bmatrix} = GJ \begin{bmatrix} 1 & -1 \\ -1 & 1 \end{bmatrix} \begin{Bmatrix} \phi_i \\ \phi_j \end{Bmatrix} \qquad (2.141)$$

or
$\{P_i\} = [k]\{U_i\}$

$\therefore [k]$ = stiffness matrix for a torque bar

$$[k] = GJ \begin{matrix} \phi_i & \phi_j \\ \begin{bmatrix} 1 & -1 \\ -1 & 1 \end{bmatrix} & \begin{matrix} \phi_i \\ \phi_j \end{matrix} \end{matrix} \qquad (2.142)$$

Examples for Practice 2

1. Determine the nodal displacements and member forces in the plane pin-jointed trusses of Figures 2.26(a) to 2.26(c). All members are of constant AE.

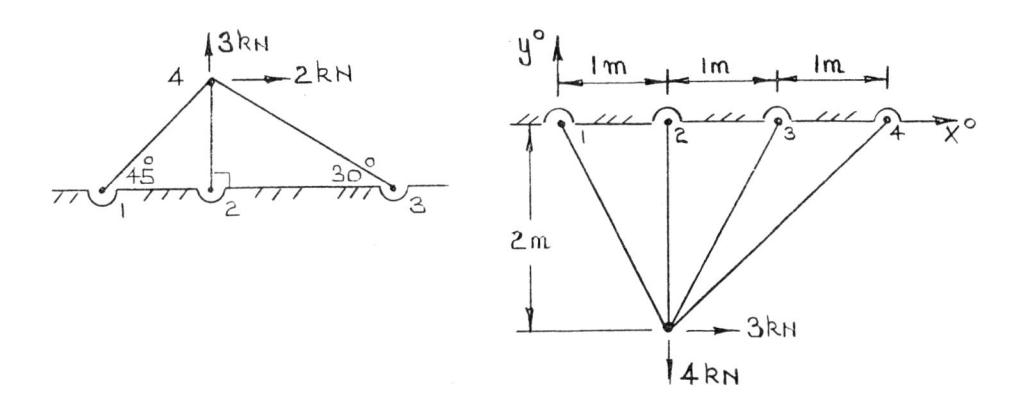

Figure 2.26(a)

Figure 2.26(b)

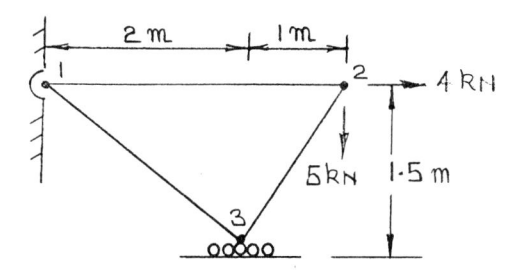

Figure 2.26(c)

2. Determine the nodal displacements and member forces for the three-dimensional pin-jointed truss of Figure 2.27.

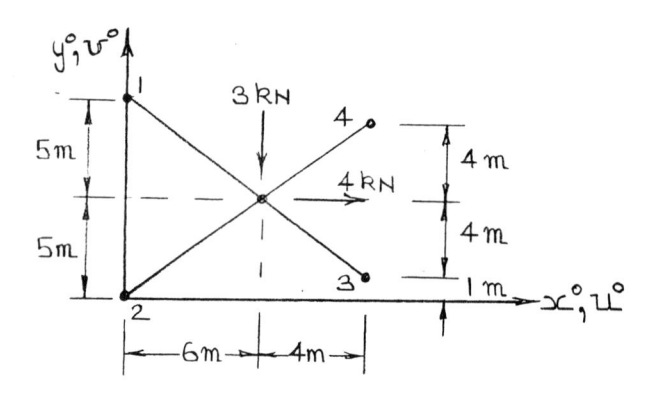

(a) Plan view

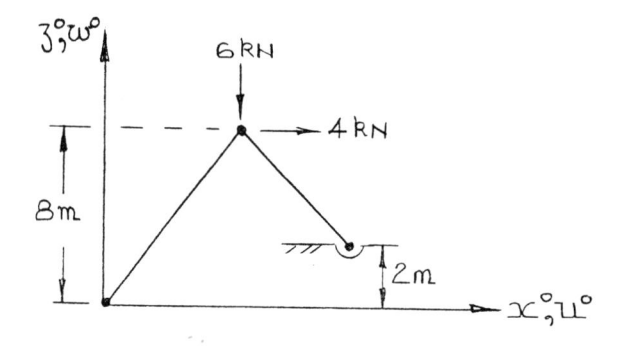

(b) Front elevation

Figure 2.27

Stiffness Matrix for a Torque Bar

3. Using the matrix displacement method, determine the nodal displacements and bending moments for the beams of Figures 2.28(a) and 2.28(b).

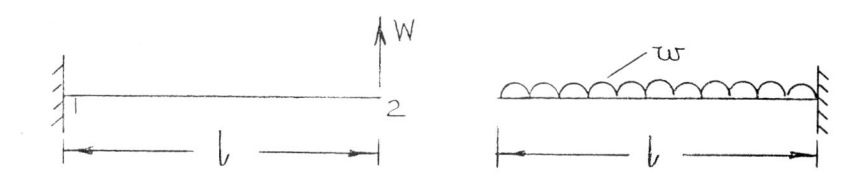

Figure 2.28(a)

Figure 2.28(b)

4. Using the matrix displacement method, determine the nodal displacements and bending moments for the continuous beams of Figures 2.29(a) and 2.29(b).

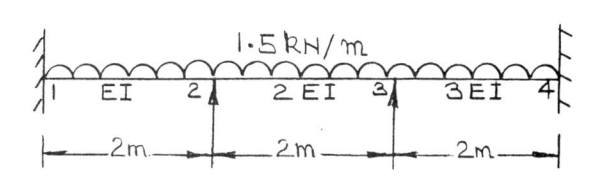

Figure 2.29(a).

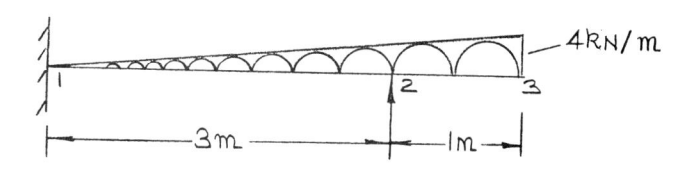

Figure 2.29(b).

It should be noted that the end fixing forces for the hydrostatically loaded beam element of Figure 2.29(c) are given by:

$R_1 = w_1 \ell/2 + (w_2 - w_1) \cdot 3\ell/20$

$R_2 = (w_1 + w_2) \ell/2 - R_1$

$M_1 = - w \ell^2/12 - (w_2 - w_1) \ell^2/30$

$M_2 = R_1 \ell + M_1 - w_1 \ell^2/2 - (w_2 - w_1)\ell^2/6$

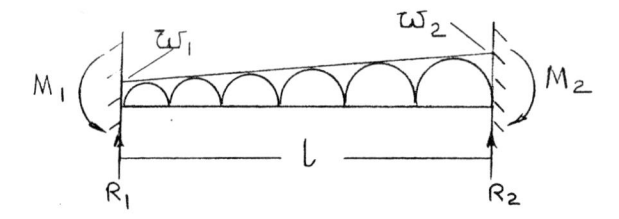

Figure 2.29(c).

5. Determine the nodal displacements and bending moments for the rigid-jointed plane frame of Figure 2.30, assuming that:

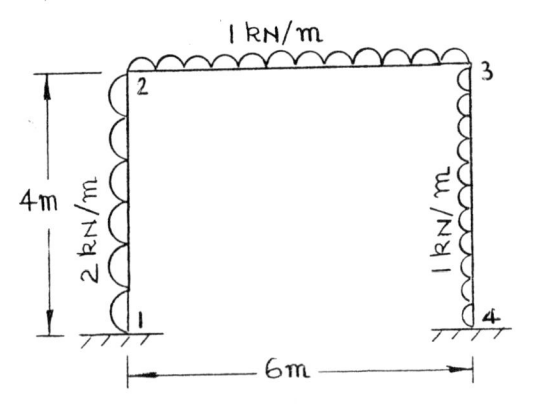

$u_1° = v_1° = \theta_1 = u_4° = v_4° = \theta_4 = v_2° = v_3° = 0$ and that $u_2° = u_3°$

Figure 2.30.

6. Determine the nodal displacements and bending moments for the rigid-jointed plane frame of Figure 2.31.

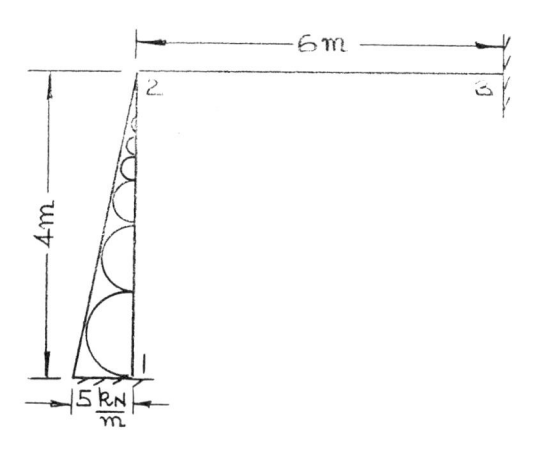

Figure 2.31.

It may be assumed that:-

E $= 2 \times 10^{11}$ N/m^2

A   = cross-sectional areas of elements

     $= 20$ cm$^2 = 0.002$ m^2

I   = second moments of area of the elements $= 1 \times 10^{-5}$ m^4.

CHAPTER 3

The Finite Element Method

3.1 Introduction

In this chapter, the finite element method proper will be introduced. The finite element method is based on the matrix displacement method, in that the unknown nodal displacements have to be calculated, prior to determining the elemental stresses.

The main problem, however, with the finite element method, is in determining the stiffness matrices of elements of complex shape. That is, the methods of Chapter 2 cannot be used to determine the stiffness matrices of triangular or quadrilateral plate elements or of thin or thick doubly curved shell elements, because these shapes are too complex for the approaches of Chapter 2.

The finite element method is particularly useful for solving complex partial differential equations which apply over a complex parent shape. This process is achieved by subdividing the complex parent shape into many finite elements of simpler shape, as shown by Figure I.1, and solving the complex partial differential equation over each of the finite elements of simpler shape. Then by considering equilibrium and compatibility at the inter element boundaries, a large number of simultaneous equations are obtained. Solution of these simultaneous equations results in values of the unknown function; in the present chapter these unknown functions are in fact displacements. Apart from its use for structural analysis, the finite element method can be used for vibrations, acoustics, electrostatics, magnetostatics, heat transfer, fluid flow, etc.

The true finite element method was invented by Turner et al [5], when they presented the three node in-plane triangular plate element. The derivation of the stiffness matrix of this element will now be described.

3.2 Stiffness matrix for the in-plane triangular element

This element is described by three corner nodes, namely nodes 1, 2 and 3, as shown in Figure 3.1.

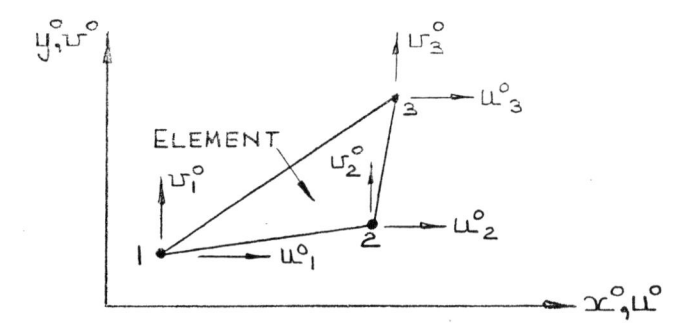

Figure 3.1 - In-plane triangular plate element.

The element is useful for mathematically modelling flat plates subjected to in-plane forces. Each node has two degrees of freedom, making a total of 6 degrees of freedom per element; these degrees of freedom, or boundary values, are u_1^o, v_1^o, u_2^o, v_2^o, u_3^o and v_3^o.

Stiffness Matrix for the In-plane Triangular Element

The method of derivation of the stiffness matrix is based on boundary values, and as there are 6 boundary values, it will be necessary to assume polynomials for the displacements u° and v°, which have a total of 6 arbitrary constant α_i, as shown by equations (3.1) and (3.2).

$$u° = \alpha_1 + \alpha_2 x° + \alpha_3 y° \tag{3.1}$$

$$v° = \alpha_4 + \alpha_5 x° + \alpha_6 y° \tag{3.2}$$

By applying each of the six boundary values to equations (3.1) and (3.2), six simultaneous equations will result, hence the 6 α's can be determined, as follows:

3.2.1 Boundary Conditions

The 6 boundary conditions are:

At $x° = x_1°$ and $y° = y_1°$, $u° = u_1°$ and $v° = v_1°$

At $x° = x_2°$ and $y° = y_2°$, $u° = u_2°$ and $v° = v_2°$ \qquad (3.3)

At $x° = x_3°$ and $y° = y_3°$, $u° = u_3°$ and $v° = v_3°$

Substituting these 6 boundary values into equations (3.1) and (3.2), the following 6 simultaneous equations are obtained:

$$u_1° = \alpha_1 + \alpha_2 x_1° + \alpha_3 y_1°$$

$$u_2° = \alpha_1 + \alpha_2 x_2° + \alpha_3 y_2° \tag{3.4}$$

$$u_3° = \alpha_1 + \alpha_2 x_3° + \alpha_3 y_3°$$

$$v_1° = \alpha_4 + \alpha_5 x_1° + \alpha_6 y_1°$$

$$v_2° = \alpha_4 + \alpha_5 x_2° + \alpha_6 y_2° \tag{3.5}$$

$$v_3° = \alpha_4 + \alpha_5 x_3° + \alpha_6 y_3°$$

Rewriting equations (3.4) and (3.5) in matrix form, the following is obtained:

$$\begin{Bmatrix} u_1° \\ u_2° \\ u_3° \\ v_1° \\ v_2° \\ v_3° \end{Bmatrix} = \left[\begin{array}{ccc|ccc} 1 & x_1° & y_1° & & & \\ 1 & x_2° & y_2° & & O_3 & \\ 1 & x_3° & y_3° & & & \\ \hline & & & 1 & x_1° & y_1° \\ & O_3 & & 1 & x_2° & y_2° \\ & & & 1 & x_3° & y_3° \end{array} \right] \begin{Bmatrix} \alpha_1 \\ \alpha_2 \\ \alpha_3 \\ \alpha_4 \\ \alpha_5 \\ \alpha_6 \end{Bmatrix} \tag{3.6}$$

$$= \begin{bmatrix} A & O_3 \\ O_3 & A \end{bmatrix} \{\alpha_i\} \qquad (3.7)$$

or $\{\alpha_i\}$
$$\begin{bmatrix} A^{-1} & O_3 \\ O_3 & A^{-1} \end{bmatrix} \{u_i^o\} \qquad (3.8)$$

where

$$[A] = \begin{bmatrix} 1 & x_1^o & y_1^o \\ 1 & x_2^o & y_2^o \\ 1 & x_3^o & y_3^o \end{bmatrix}$$

&

$$[A^{-1}] = \begin{bmatrix} a_1 & a_2 & a_3 \\ b_1 & b_2 & b_3 \\ c_1 & c_2 & c_3 \end{bmatrix} / \det |A| \qquad (3.9)$$

$$a_1 = x_2 y_3^o - x_3^o y_2^o$$
$$a_2 = x_3^o y_1^o - x_1 y_3^o \qquad (3.10)$$
$$a_3 = x_1^o y_2^o - x_2^o y_1^o$$

$$b_1 = y_2^o - y_3^o$$
$$b_2 = y_3^o - y_1^o \qquad (3.11)$$
$$b_3 = y_1^o - y_2^o$$

$$c_1 = x_3^o - x_2^o$$
$$c_2 = x_1^o - x_3^o \qquad (3.12)$$
$$c_3 = x_2^o - x_1^o$$

$$\det |A| = x_2^o y_3^o - y_2^o y_3^o - x_1^o (y_3^o - y_2^o) + y_1^o (x_3^o - x_2^o)$$
$$= 2\Delta$$

Δ = area of triangular element

Hence,

$$u^\circ = N_1 u_1^\circ + N_2 u_2^\circ + N_3 u_3^\circ$$

and

$$v^\circ = N_1 v_1^\circ + N_2 v_2^\circ + N_3 v_3^\circ \tag{3.13}$$

or

$$\begin{Bmatrix} u^\circ \\ v^\circ \end{Bmatrix} = \begin{bmatrix} N_1 & N_2 & N_3 & 0 & 0 & 0 \\ 0 & 0 & 0 & N_1 & N_2 & N_3 \end{bmatrix} \begin{Bmatrix} u_1^\circ \\ u_2^\circ \\ u_3^\circ \\ v_1^\circ \\ v_2^\circ \\ v_3^\circ \end{Bmatrix} \tag{3.14}$$

$$= [N]\{U_i\}$$

where

[N] = a matrix of shape functions

$$= \begin{bmatrix} N_1 & N_2 & N_3 & 0 & 0 & 0 \\ 0 & 0 & 0 & N_1 & N_2 & N_3 \end{bmatrix} \tag{3.15}$$

$$N_1 = \frac{1}{2\Delta} (a_1 + b_1 x^\circ + c_1 y^\circ)$$

$$N_2 = \frac{1}{2\Delta} (a_2 + b_2 x^\circ + c_2 y^\circ) \tag{3.16}$$

$$N_3 = \frac{1}{2\Delta} (a_3 + b_3 x^\circ + c_2 y^\circ)$$

3.2.2 The matrix [B]

Now for a two dimensional system of stress, the strains [6] are given by:

$$\varepsilon_x = \partial u^\circ / \partial x^\circ = \text{strain in the } x^\circ \text{ direction}$$

$$\varepsilon_y = \partial v^\circ / \partial y^\circ = \text{strain in the } y^\circ \text{ direction} \tag{3.17}$$

$$\gamma_{xy} = \partial u^\circ / \partial y^\circ + \partial v^\circ / \partial x^\circ = \text{shear strain in the } x^\circ \text{ - } y^\circ \text{ plane}$$

Hence, from equations (3.13) and (3.17)

$$\varepsilon_x = \frac{1}{2\Delta} (b_1 u_1^\circ + b_2 u_2^\circ + b_3 u_3^\circ)$$

$$\varepsilon_y = \frac{1}{2\Delta} (c_1 v_1^\circ + c_2 v_2^\circ + c_3 v_3^\circ) \qquad (3.18)$$

$$\gamma_{xy} = \frac{1}{2\Delta} (c_1 u_1^\circ + c_2 u_2^\circ + c_3 u_3^\circ + b_1 v_1^\circ + b_2 v_2^\circ + b_3 v_3^\circ)$$

or in the matrix form

$$\begin{Bmatrix} \varepsilon_x \\ \varepsilon_y \\ \gamma_{xy} \end{Bmatrix} = \frac{1}{2\Delta} \begin{bmatrix} b_1 & b_2 & b_3 & 0 & 0 & 0 \\ 0 & 0 & 0 & c_1 & c_2 & c_3 \\ c_1 & c_2 & c_3 & b_1 & b_2 & b_3 \end{bmatrix} \begin{Bmatrix} u_1^\circ \\ u_2^\circ \\ u_3^\circ \\ v_1^\circ \\ v_2^\circ \\ v_3^\circ \end{Bmatrix} \qquad (3.19)$$

$$= [B] \{u_i\}$$

where

[B] = a matrix relating co-ordinate strains to nodal displacements

$$= \frac{1}{2\Delta} \begin{bmatrix} b_1 & b_2 & b_3 & 0 & 0 & 0 \\ 0 & 0 & 0 & c_1 & c_2 & c_3 \\ c_1 & c_2 & c_3 & b_1 & b_2 & b_3 \end{bmatrix} \qquad (3.20)$$

3.2.3 The Matrix [D]

Now the stress-strain relationships for an in-plane plate, in **plane stress** [4], are given by:

$$\sigma_x = \frac{E}{(1 - v^2)} (\varepsilon_x + v \varepsilon_y)$$

$$\sigma_y = \frac{E}{(1 - v^2)} (\varepsilon_y + v \varepsilon_x) \qquad (3.21)$$

$$\tau_{xy} = G \gamma_{xy}$$

where,

σ_x = normal stress in x° direction

σ_y = normal stress in y° direction

τ_{xy} = shear stress in the x° - y° plane

E = Young's modulus

G = Rigidity modulus = E/[2 (1 + v)]

v = Poisson's ratio

In matrix form, equations(3.21) appear as:-

$$\begin{Bmatrix} \sigma_x \\ \sigma_y \\ \tau_{xy} \end{Bmatrix} = \frac{E}{(1 - v^2)} \begin{bmatrix} 1 & v & 0 \\ v & 1 & 0 \\ 0 & 0 & (1 - v)/2 \end{bmatrix} \begin{Bmatrix} \epsilon_x \\ \epsilon_y \\ \gamma_{xy} \end{Bmatrix} \quad (3.22)$$

or $\{\sigma\} = [D]\{\epsilon\}$ (3.23)

where

[D] = a matrix of material constants

$$= \frac{E}{(1 - v^2)} \begin{bmatrix} 1 & v & 0 \\ v & 1 & 0 \\ 0 & 0 & (1 - v)/2 \end{bmatrix} \quad (3.24)$$

For **plane strain** [4], which is a two dimensional system of strain and a three dimensional system of stress, the matrix of material constants [D], is given by:-

$$[D] = \frac{E}{(1 + v)(1 - 2v)} \begin{bmatrix} (1 - v) & v & 0 \\ v & (1 - v) & 0 \\ 0 & 0 & (1 - 2v)/2 \end{bmatrix} \quad (3.25)$$

In general

$$[D] = E^1 \begin{bmatrix} 1 & \mu & 0 \\ \mu & 1 & 0 \\ 0 & 0 & \gamma \end{bmatrix} \quad (3.26)$$

where for **plane stress**

$E^1 = E/(1 - v^2)$

$\mu = v$ (3.27)

$\gamma = (1 - v)/2$

and for **plane strain**

$$E^1 = \frac{E}{(1 + v)(1 - 2v)}$$

$\mu = v$ (3.28)

$\gamma = (1 - 2v)/2$

3.2.4 The stiffness matrix

Now from reference [4], the elastic strain energy stored in a body =

$$U_e = 1/2E \int \sigma^2 \, d(\text{vol}) \qquad (3.29)$$

but $\dfrac{\sigma}{\epsilon} = E$

or $\sigma = E \epsilon$

$\therefore U_e = \tfrac{1}{2} \int E \, \epsilon^2 \, d(\text{vol})$ (3.30)

In matrix form, equation (3.30) appears as

$$U_e = \tfrac{1}{2} \int \{\epsilon\}^T [D] \{\epsilon\} \, d(\text{vol}) \qquad (3.31)$$

It is necessary to write equation (3.31) in the form shown, because if it is multiplied out, the result will be a scalar, which is the correct form for strain energy.

but $[\epsilon] = [B] \{U_i\}$ (3.32)

$\therefore U_e = [U_i]^T \int [B]^T [D] [B] \{U_i\} \, d(\text{vol})$ (3.33)

However, the potential of the external loads $= - \{U_i\}^T \{P_i\}$ (3.34)

and the total potential is given by:

$\pi = U_e - \{U_i\}^T \{P_i\}$ (3.35)

For minimum potential,

$$\frac{d\pi_p}{d\{U_i\}} = 0 \qquad (3.36)$$

$$\therefore \{P_i\} = \int [B]^T [D] [B] \, d(\text{vol}) \{U_i\} \qquad (3.37)$$

$$= [k] \{U_i\}$$

$\therefore [k°]$ = elemental stiffness matrix in global co-ordinates

$$= \int [B]^T [D] [B] \, d(\text{vol}) \qquad (3.38)$$

Substituting equations (3.20) and (3.26) into equation (3.38)

$$[k°] = \int \frac{1}{4\Delta^2} \begin{bmatrix} b_1 & 0 & c_1 \\ b_2 & 0 & c_2 \\ b_3 & 0 & c_3 \\ 0 & c_1 & b_1 \\ 0 & c_2 & b_2 \\ 0 & c_3 & b_3 \end{bmatrix} E^1 \begin{bmatrix} 1 & \mu & 0 \\ \mu & 1 & 0 \\ 0 & 0 & \gamma \end{bmatrix} \begin{bmatrix} b_1 & b_2 & b_3 & 0 & 0 & 0 \\ 0 & 0 & 0 & c_1 & c_2 & c_3 \\ c_1 & c_2 & c_3 & b_1 & b_2 & b_3 \end{bmatrix} t \, dA \qquad (3.39)$$

As b_1, b_2, b_3 etc are constants, $\int dA = \Delta$

$$\therefore [k°] = t \begin{bmatrix} P_{ij} & Q_{ij} \\ \hline Q_{ji} & R_{ij} \end{bmatrix} \qquad (3.40)$$

where

t = plate thickness

$P_{ij} = 0.25 \, E^1 \, (b_i \, b_j + \gamma \, c_i \, c_j)/\Delta$

$Q_{ij} = 0.25 \, E^1 \, (\mu b_i \, c_j + \gamma \, c_i \, b_j)/\Delta$

$Q_{ji} = 0.25 \, E^1 \, (\mu b_j \, c_i + \gamma \, c_j \, b_i)/\Delta$ $\qquad (3.41)$

$R_{ij} = 0.25 \, E^1 \, (c_i \, c_j + \gamma \, b_i \, b_j)/\Delta$

where i and j vary from 1 to 3.

NB It should be noted that $[k°]$ is of order 6, which is the same number as the degrees of freedom for the element.

3.3 Stiffness matrix for an in-plane annular plate

The in-plane annular plate is assumed to deform axisymmetrically. That is, any circumference in the plate is assumed to remain of circumferential form during deformation. Because of this assumption, it will be convenient to assume that the in-plane annular element is governed by two ring nodes, as shown in Figure 3.2. Additionally, it must be emphasised that the annular plate is subjected only to in-plane loads, so that it is either in tension or compression, and that no flexure occurs.

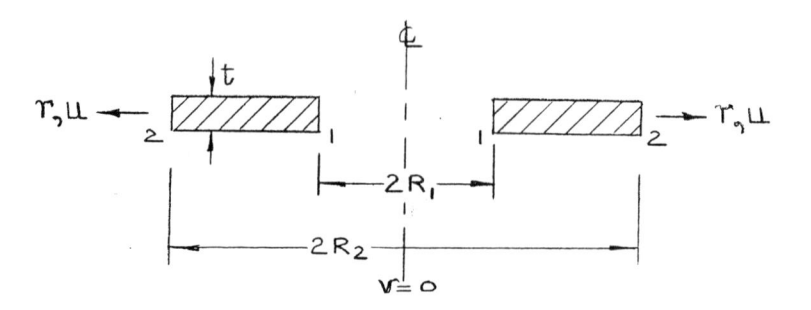

Figure 3.2 - In-plane annular element.

In this case there are two degrees of freedom, namely u_1 and u_2, which occur at nodal circles 1 and 2, respectively. As there are two degrees of freedom, it will be convenient to assume a polynomial with two arbitrary constants, namely α_1 and α_2, as shown by equation (3.42). It should be emphasised that it is necessary to assume the same number of α_i's as the boundary values so that the correct number of simultaneous equations are obtained

$$u = \alpha_1 + \alpha_2 r \tag{3.42}$$

3.3.1 To obtain [N], the matrix of shape functions

The boundary values are:

at $r = R_1$, $u = u_1$

and

at $r = R_2$, $u = u_2$

Substituting these boundary values into equation (3.42), the following two simultaneous equations are obtained:-

$$u_1 = \alpha_1 + \alpha_2 R_1 \tag{3.43}$$

and

$$u_2 = \alpha_1 + \alpha_2 R_2 \tag{3.44}$$

Taking equation (3.43) away from equation (3.44) will eliminate α_1, as follows, so that an expression for α_2 can be obtained

$$u_2 - u_1 = \alpha_2 (R_2 - R_1)$$

$$\text{or } \alpha_2 = \frac{(u_2 - u_1)}{(R_2 - R_1)} \tag{3.45}$$

Substituting equation (3.45) into equation (3.43)

$$u_1 = \alpha_1 + (u_2 - u_1) \frac{R_1}{(R_2 - R_2)}$$

$$\text{or } \alpha_1 = u_1 - (u_2 - u_1) \frac{R_1}{(R_2 - R_1)}$$

$$= \frac{u_1 (R_2 - R_1) - (u_2 - u_1) R_1}{(R_2 - R_1)}$$

$$= \frac{u_1 (R_2 - R_1 + R_1) - u_2 R_1}{(R_2 - R_1)}$$

$$\alpha_1 = \frac{u_1 R_2 - u_2 R_1}{(R_2 - R_1)} \tag{3.46}$$

Substituting equations (3.45) and (3.46) into equation (3.42)

$$u = \frac{u_1 R_2 - u_2 R_1}{(R_2 - R_1)} + \frac{(u_2 - u_1)}{(R_2 - R_1)} r$$

$$= \frac{u_1 (R_2 - r) + u_2 (-R_1 + r)}{(R_2 - R_1)}$$

$$= \frac{1}{(R_2 - R_1)} [(R_2 - r) \; (-R_1 + r)] \begin{Bmatrix} u_1 \\ u_2 \end{Bmatrix} \tag{3.47}$$

$$u = [N] \{U_i\}$$

or [N] = the matrix of shape functions

$$[N] = [(R_2 - r) \; (-R_1 + r)]/(R_2 - R_1) \tag{3.48}$$

3.3.2 To obtain [B]

Now for the in-plane annular element [4] the radial and hoop strains are principal strains, and are given by:

$$\varepsilon_r = \text{radial strain } \frac{du}{dr}$$

$$= [u_1(-1) + u_2(1)]/(R_2 - R_1)$$

$$= [-1 \quad 1]/(R_2 - R_1) \begin{Bmatrix} u_1 \\ u_2 \end{Bmatrix}$$

ε_ϕ = hoop or circumferential strain = u/r

$$= \frac{1}{(R_2 - R_1)} [(R_2 - r)/r \quad (-R_1 + r)/r] \begin{Bmatrix} u_1 \\ u_2 \end{Bmatrix}$$

$$\text{or } \{\varepsilon\} = \begin{Bmatrix} \varepsilon_r \\ \varepsilon_\phi \end{Bmatrix} = \frac{1}{(R_2 - R_1)} \begin{bmatrix} -1 & 1 \\ (R_2 - r)/r & (-R_1 + r)/r \end{bmatrix} \begin{Bmatrix} u_1 \\ u_2 \end{Bmatrix} \quad (3.49)$$

$$\text{ie } [B] = \frac{1}{(R_2 - R_1)} \begin{bmatrix} -1 & 1 \\ (R_2 - r)/r & (-R_1 + r)/r \end{bmatrix}$$

NB $\gamma_{r\phi} = 0$

3.3.3 To obtain [D]

Now as ε_r and ε_ϕ are principal strains, the stress-strain relationships are:-

$\sigma_r = E^1 (\varepsilon_r + \mu\varepsilon_\phi)$
$\sigma_\phi = E^1 (\mu\varepsilon_r + \varepsilon_\phi)$
$\tau_{r\phi} = 0$

or in matrix form

$$\{\sigma\} = \begin{Bmatrix} \sigma_r \\ \sigma_\phi \end{Bmatrix} = E^1 \begin{bmatrix} 1 & \mu \\ \mu & 1 \end{bmatrix} \begin{Bmatrix} \varepsilon_r \\ \varepsilon_\phi \end{Bmatrix}$$

$$\text{or } [D] = E^1 \begin{bmatrix} 1 & \mu \\ \mu & 1 \end{bmatrix} \quad (3.50)$$

3.3.4 To obtain [k]

Substituting equations (3.49) and (3.50) into equation (3.38).

$$[k] = \int_{R_1}^{R_2} \frac{1}{(R_2 - R_1)^2} \begin{bmatrix} -1 & (R_2-r)/r \\ 1 & (r-R_1)/r \end{bmatrix} E^1 \begin{bmatrix} 1 & \mu \\ \mu & 1 \end{bmatrix}$$

$$\begin{bmatrix} -1 & 1 \\ \dfrac{(R_2-r)}{r} & \dfrac{(r-R_1)}{r} \end{bmatrix} 2\pi r \, dr \, t$$

$$= \begin{bmatrix} k_{11} & k_{12} \\ k_{21} & k_{22} \end{bmatrix} \quad (3.51)$$

where

$k_{11} = CN \{R_2[R_2 \ln (R_2/R_1) - 2(1+\mu)(R_2 - R_1)]$

$+ (1+\mu)(R_2^2 - R_1^2)\}$

$k_{12} = k_{21} = CN [-R_1 R_2 \ln (R_2/R_1)]$

$k_{22} = CN \{R_1 [R_1 \ln (R_2/R_1) - 2(1+\mu)(R_2 - R_1)]$ \quad (3.52)

$+ (1+\mu)(R_2^2 - R_1^2)$

$CN = 2\pi E^1 t/(R_2 - R_1)^2$

NB It should be noted that the order of [k] is 2, which is the same number as the degrees of freedom of the element.

3.4 Three node rod element

This element, which is of uniform cross-section, has a mid-side node in addition to the end nodes, as shown in Figure 3.3. Very often, a better stiffness matrix can be obtained if additional mid-side nodes are used.

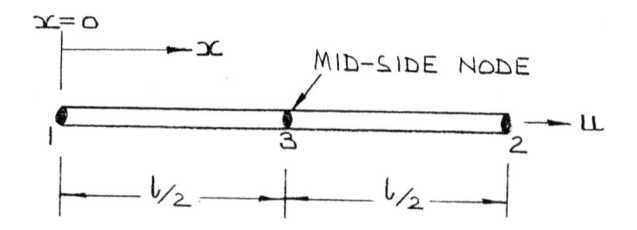

Figure 3.3 Rod element with 3 nodes.

3.4.1 To obtain [N]

In this case, there are 3 nodal displacements, namely u_1, u_2 and u_3, hence, it will be necessary to assume a polynomial with 3 α_i's as shown by equation (3.53).

$$u = \alpha_1 + \alpha_2 x + \alpha_3 x^2 \tag{3.53}$$

The three boundary values are:

at $x = 0$, $u = u_1$

at $x = \ell$, $u = u_2$ \hfill (3.54)

and

at $x = \ell/2$, $u = u_3$

Substituting these three boundary values into equation (3.53), the following three simultaneous equations are obtained:

$$u_1 = \alpha_1$$

$$u_2 = \alpha_1 + \alpha_2 \ell + \alpha_3 \ell^2 \tag{3.55}$$

$$u_3 = \alpha_1 + \alpha_2 \ell/2 + \alpha_3 \ell^2/4$$

ie $\underline{\alpha_1 = u_1}$ \hfill (3.56)

Hence, from equations (3.55b) and (3.55c)

$$u_2 - u_1 = \alpha_2 \ell + \alpha_3 \ell^2 \tag{3.57}$$

$$u_3 - u_1 = \alpha_2 \ell/2 + \alpha_3 \ell^2/4 \tag{3.58}$$

To eliminate α_2, multiply equation (3.58) by 2 and take it away from equation (3.57) to give

$$u_2 - u_1 - 2u_3 + 2u_1 = \alpha_3 \ell^2 (1 - 1/2)$$

or $\alpha_3 \ell^2/2 = u_1 + u_2 - 2u_3$

or $\alpha_3 = (2u_1 + 2u_2 - 4u_3)/\ell^2$ \hfill (3.59)

Substituting equation (3.59) into equation (3.57).

$u_2 - u_1 = \alpha_2 \ell + 2u_1 + 2u_2 - 4u_3$

$\alpha_2 = (u_2 - u_1 - 2u_1 - 2u_2 + 4u_3)/\ell$

$\alpha_2 = (-3u_1 - u_2 + 4u_3)/\ell$ \hfill (3.60)

Substituting equations (3.56), (3.59) and (3.60) into equation (3.53).

$u = u_1 + (-3u_1 - u_2 + 4u_3) \, x/\ell$

$+ (2u_1 + 2u_2 - 4u_3) x^2/\ell^2$ \hfill (3.61)

Let, $\xi = x/\ell$

$\therefore u = u_1 (-3u_1 - u_2 + 4u_3) \xi + (2u_1 + 2u_2 - 4u_3)\xi^2$ \hfill (3.62)

$= u_1 (1 - 3\xi + 2\xi^2) + u_2 (-\xi + 2\xi^2)$

$+ u_3 (4\xi - 4\xi^2)$

$u = [(1 - 3\xi + 2\xi^2) \; (-\xi + 2\xi^2) \; (4\xi - 4\xi^2)] \begin{Bmatrix} u_1 \\ u_2 \\ u_3 \end{Bmatrix}$ \hfill (3.63)

$= [N] \{U_i\}$

$\therefore [N]$ = the matrix of shape functions

$= [(1 - 3\xi + 2\xi^2) \; (-\xi + 2\xi^2) \; (4\xi - 4\xi^2)]$ \hfill (3.64)

3.4.2 To obtain [D]

In one dimension, and from Hooke's law,

$\sigma = E\varepsilon$

or in matrix form,

$\{\sigma\} = [D] \{\varepsilon\}$

$\therefore [D] = [E]$ \hfill (3.65)

3.4.3 To obtain [B]

In one dimension,

ε = axial strain

$$\varepsilon = \frac{du}{dx} = \frac{du}{\ell d\zeta}$$

$$= [u_1 (-3 + 4\xi) + u_2 (-1 + 4\xi) + u_3 (4 - 8\xi)]/\ell$$

$$= [(-3 + 4\xi) \ (-1 + 4\xi) \ (4 - 8\xi)] \begin{Bmatrix} u_1 \\ u_2 \\ u_3 \end{Bmatrix} / \ell$$

or

$$u = [B] \{U_i\}$$

$$\therefore [B] = [(-3 + 4\xi) \ (-1 + 4\xi) \ (4 - 8\xi)]/\ell \qquad (3.66)$$

Substituting equations (3.65) and (3.66) into equation (3.38),

$[k]$ = the stiffness matrix for a 3 node rod element

$$= \tfrac{1}{2} \int_0^1 \frac{1}{\ell^2} \begin{bmatrix} (-3 + 4\xi) \\ (-1 + 4\xi) \\ (4 - 8\xi) \end{bmatrix} E[(-3 + 4\xi) \ (-1 + 4\xi) \ (4 - 8\xi)]$$

$A.\ell d\xi$

$$[k] = \begin{bmatrix} k_{11} & k_{12} & k_{13} \\ k_{21} & k_{22} & k_{23} \\ k_{31} & k_{32} & k_{33} \end{bmatrix} \qquad (3.67)$$

where

$$[k_{11}] = \frac{AE}{\ell} \int_0^1 (-3 + 4\xi)^2 \, d\xi$$

$$= \frac{AE}{\ell} \int_0^1 (9 - 24\xi + 16\xi^2) d\xi \qquad (3.68)$$

$$= \frac{AE}{\ell} (9 - 12 + \frac{16}{3}) = 7 \, AE/(3\ell)$$

$$k_{12} = k_{21} = \frac{AE}{\ell} \int_0^1 (-3 + 4\xi)(-1 + 4\xi) \, d\xi$$

$$= \frac{AE}{\ell} \int_0^1 (3 - 16\xi + 16\xi^2) \, d\xi \qquad (3.69)$$

$$= \frac{AE}{\ell} (3 - \frac{16}{2} + \frac{16}{3})$$

$$= AE/(3\ell)$$

$$k_{13} = k_{31} = \frac{AE}{\ell} \int_0^1 (-3 + 4\xi)(4 - 8\xi) \, d\xi$$

$$= \frac{AE}{\ell} \int_0^1 (-12 + 40\xi - 32\xi^2) \, d\xi \qquad (3.70)$$

$$= \frac{AE}{\ell} (-12 + \frac{40}{2} - \frac{32}{3})$$

$$= -8AE/(3\ell)$$

$$k_{22} = \frac{AE}{\ell} \int_0^1 (-1 + 4\xi)(-1 + 4\xi) \, d\xi$$

$$= \frac{AE}{\ell} \int_0^1 (1 - 8\xi + 16\xi^2) \, d\xi \qquad (3.71)$$

$$= \frac{AE}{\ell} (1 - \frac{8}{2} + \frac{16}{3})$$

$$= 7AE/(3\ell)$$

$$k_{23} = k_{32} = \frac{AE}{\ell} \int_0^1 (-1 + 4\xi)(4 - 8\xi) \, d\xi$$

$$= \frac{AE}{\ell} \int_0^1 (-4 + 24\xi - 32\xi^2) \, d\xi \qquad (3.72)$$

$$= \frac{AE}{\ell} \left(-4 + \frac{24}{2} - \frac{32}{3} \right)$$

$$= -8AE/(3\ell)$$

$$k_{33} = \frac{AE}{\ell} \int_0^1 (4 - 8\xi)(4 - 8\xi) \, d\xi$$

$$= \frac{AE}{\ell} \int (16 - 64\xi + 64\xi^2) \, d\xi \qquad (3.73)$$

$$= \frac{AE}{\ell} \left(16 - \frac{64}{2} + \frac{64}{3} \right)$$

$$= 16AE/(3\ell)$$

NB It should be noted that [k] is order 3, which is the same number as the degrees of freedom of the element.

3.5 Distributed Loads

In the finite element method, it is necessary to apply the loads at the nodal points. Hence, it is necessary to represent distributed loads as equivalent nodal loads; this is achieved as follows:-

Let $\{P_w\}$ = a vector of nodal loads, equivalent to the distributed load acting on the element.

The potential of the nodal loads, equivalent to the distributed load

$$= - \{U_i\}^T \{P_w\} \qquad (3.74)$$

If the distributed load is a function of x, y and z, then the potential of the distributed load on any element

$$= - \int \{u\}^T [P_{(x)}] \, d\,(vol) \qquad (3.75)$$

but $\{u\}^T = \{U_i\}^T [N]^T \qquad (3.76)$

Distributed Loads

∴ potential of distributed load

$$= -\{U_i\}^T \int [N]^T [P_{(x)}] \, d(\text{vol}) \tag{3.77}$$

Equating (3.74) and (3.77)

$$\{P_w\} = \int [N]^T [P_{(x)}] \, d(\text{vol}) \tag{3.78}$$

3.5.1 For a uniformly distributed load of value w, acting on a beam of length ℓ

$$[P_{(x)}] = w$$

$$[N]^T = \begin{bmatrix} (1 - 3\xi^2 + 2\xi^3) \\ \ell(-\xi + 2\xi^2 - \xi^3) \\ (3\xi^2 - 2\xi^3) \\ \ell(2\xi^2 - \xi^3) \end{bmatrix} \tag{3.79}$$

$d(\text{vol}) = dx = \ell \, d\xi$

where $\xi = x/\ell$

$$\therefore \{P_w\} = \int_0^1 \begin{bmatrix} (1 - 3\xi^2 + 2\xi^3) \\ \ell(-\xi + 2\xi^2 - \xi^3) \\ (3\xi^3 - 2\xi^3) \\ \ell(2\xi^2 - \xi^3) \end{bmatrix} w\ell \, d\xi$$

$$= \begin{Bmatrix} w\ell/2 \\ -w\ell^2/12 \\ w\ell/2 \\ w\ell^2/12 \end{Bmatrix} \tag{3.80}$$

Equation (3.80) can be seen to be equivalent to applying the negative resultants of the end fixing forces, as carried out in Chapter 2, and as shown in Figure 3. It must be remembered that when the stresses are calculated, the body forces, or end fixing forces, as shown in Figure 3.5, added to the nodal "stress" values.

Figure 3.4 - Equivalent nodal loads.

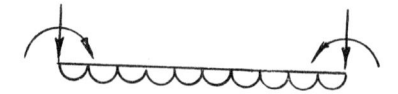

Figure 3.5 - End fixing forces.

3.6 von Mises Stress (σ_{vm})

Most finite element computer programs produce large numbers of different types of stress, but one type of stress which is probably the most important of these stresses, is called the von Mises stress (σ_{vm}), which is defined as follows:

$$\sigma_{vm} = \sqrt{(\sigma_1 - \sigma_2)^2 + (\sigma_1 - \sigma_3)^2 + (\sigma_2 - \sigma_3)^2}/\sqrt{2}$$

or in two dimensions

$$\sigma_{vm} = \sqrt{\sigma_1^2 + \sigma_2^2 - \sigma_1\sigma_2}$$

where σ_1, σ_2 and σ_3 are the principal stresses at a point in the material. The importance of the von Mises stress is that it gives a better indication of the onset of yield for two and three dimensional systems of stress than does the magnitude of the principal stresses.

That is, according to the von Mises yield criterion, the material is said to yield when the maximum value of the von Mises stresses reaches yield.

For more information on von Mises stresses, see reference [3 and 4].

NB. For composites the Azzi-Tsai stress is equivalent to the von Mises stress.

Examples for Practice 3

1. Using the finite element method, determine the elemental stiffness matrix for a uniform section rod, of length ℓ and with end nodes.

2. Determine the stiffness matrix for a rod element whose cross-sectional area varies uniformly from A_1, at node 1 to A_2 at node 2.

3. Determine the elemental stiffness matrix for a tapered rod element, of length ℓ, and with a mid-side node, namely node 3, in addition to the two end nodes. The cross-section area of the rod element may be assumed to vary linearly from A at the left end of 3A at the right end.

4. Using the finite element method, determine the stiffness matrix of a uniform section horizontal beam, of length ℓ, and with two end nodes.

5. Using finite element method, determine the stiffness matrix of a uniform section torque bar, of length ℓ, and with two end nodes.

Chapter 4

Vibration of Structures

4.1 Introduction

Mechanical vibrations are of much importance in a number of different branches of engineering, where in general, they are an undesirable phenomenon. For example, in naval architecture, such vibrations occur in ships and submarines, and in aeronautical engineering they occur in aircraft and spacecraft. In mechanical engineering, undesirable vibrations can occur in machinery and in automobiles, while in civil engineering, such vibrations can occur in tall buildings and bridges. Even in electrical engineering, undesirable vibrations can occur in power lines and in printed circuit boards.

Mechanical vibrations usually occur when a periodic force applied to a structure, has the same frequency as the natural frequency of the structure. When this occurs the structure can resonate quite dangerously, and cause severe damage to the structure apart from causing unwanted noise. The periodic force can be caused by out-of-balance rotating or reciprocating machinery, or by the motion of fluid flowing past the structure or the mechanism.

In some cases, however, mechanical vibrations are used for a positive purpose. For example, vibrations are very important in music; also the principle of the acoustic strain gauge is based on the vibrating characteristics of a stretched wire, and the mixing of large quantities of concrete is often aided with mechanical vibrators.

In this Chapter, it will be assumed that the reader is familiar with the elementary mathematical theory of mechanical vibrations. Elemental mass matrices will be derived, and the finite element method will be applied to a number of different types of structure.

4.2 The Elemental Mass Matrix

In Chapter 3, the elemental stiffness matrix [k], was derived using energy principles, and the same process will be used here to derive an expression for the elemental mass matrix [m]. However, in this case, it will only be necessary to consider kinetic energy.

Now,

$$\text{Kinetic energy} = \text{KE} = \frac{M\dot{u}^2}{2} \quad (4.1)$$

where

$$M = \text{mass}$$

$$\dot{u} = \text{velocity}$$

In matrix form,

$$\text{KE} = \frac{1}{2} \{\dot{U}_i\}^T [m] \{\dot{U}_i\} \quad (4.2)$$

where [m] = elemental mass matrix

Equation (4.2) must be written in the manner shown, because when it is multiplied out, a scalar value results and kinetic energy is, of course, a scalar quantity.

Assuming simple harmonic motion occurs it will be convenient to assume that

$$\{U_i\} = \{C\} e^{j\omega t}$$

$$\text{and } \{\dot{U}_i\} = j\omega \{C\} e^{j\omega t} \tag{4.3}$$

$$\text{where } j = \sqrt{-1}$$

$$\{C\} = \text{a vector of constants}$$

$$\text{or } \{\dot{U}_i\} = j\omega \{U_i\} \tag{4.4}$$

Substituting equation (4.4) into equation (4.2)

$$KE = -\frac{1}{2}\omega^2 \{U_i\}^T [m] \{U_i\} \tag{4.5}$$

Now for an infinitesimally small element of mass $\rho.d(vol)$,

$$KE = \frac{1}{2} \dot{u} \, \rho.d(vol) \, \dot{u}$$

and for the whole element

$$KE = \frac{1}{2} \int_{vol} \dot{u} \, \rho \, \dot{u} \, d(vol) \tag{4.6}$$

but from equation (3.14)

$$u = [N] \{U_i\} \tag{4.7}$$

$$\text{and } \dot{u} = [N] \{\dot{U}_i\} \tag{4.8}$$

Substituting equation (4.4) into equation (4.8)

$$\dot{u} = j\omega [N] \{U_i\} \tag{4.9}$$

Substituting equation (4.9) into equation (4.6)

$$KE = \frac{-1}{2} \omega^2 \{U_i\}^T \int_{vol} [N]^T \rho [N] \, d(vol) \{U_i\} \tag{4.10}$$

Equating (4.5) and (4.10)

$$[m] = \text{elemental mass matrix}$$
$$= \int_{vol} [N]^T \rho [N] \, d(vol) \qquad (4.11)$$

4.3 Mass Matrix for a Rod Element

The one dimensional rod element is shown in Figure 4.1.

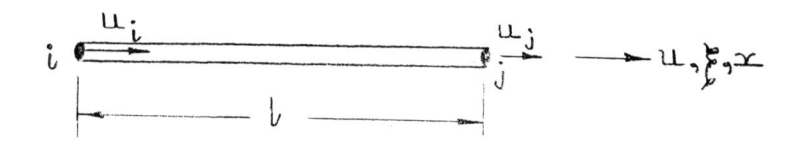

Figure 4.1 One dimensional rod element.

4.3.1 To obtain [N]

From Figure 4.1, the one dimensional rod element has two degrees of freedom, namely u_i and u_j, hence, two constants must be assumed for the polynomial describing the displacement "u", as follows

$$u = \alpha_1 + \alpha_2 x \qquad (4.12)$$

To obtain [N], it will be necessary to obtain α_1 and α_2 in terms of the boundary values namely u_i and u_j

$$\text{At } x = 0, \; u = u_i \qquad (4.13)$$

Substituting the boundary value from equation (4.13) into equation (4.12)

$$u_i = \alpha_1$$
$$\text{or } \alpha_1 = u_i \qquad (4.14)$$

Mass Matrix for a Rod Element

The second boundary value is at

$$x = \ell, \quad u = u_j \tag{4.15}$$

Substituting equation (4.15) into equation (4.12)

$$u_j = \alpha_1 + \alpha_2 \ell$$

$$\text{or } \alpha_2 = (u_j - u_i)/\ell \tag{4.16}$$

Substituting equations (4.14) and (4.16) into equation (4.12)

$$u = u_i + (u_j - u_i) \, x/\ell \tag{4.17}$$

Let,

$$\xi = x/\ell$$

$$\therefore u = u_i + (u_j - u_i) \, \xi$$

$$\text{or } u = u_i(1 - \xi) + u_j \, \xi \tag{4.18}$$

$$= [(1 - \xi) \; \xi] \begin{Bmatrix} u_i \\ u_j \end{Bmatrix}$$

$$= [N] \{U_i\} \tag{4.19}$$

$$\therefore [N] = [(1 - \xi) \; \xi]$$

4.3.2 To obtain [m]

Substituting equation (4.19) into equation (4.11)

$$[m] = \int_0^1 \begin{bmatrix} (1 - \xi) \\ \xi \end{bmatrix} \rho \, [(1 - \xi) \; \xi] \, A.\ell.d\xi$$

$$= \rho A \ell \int_0^1 \begin{bmatrix} (1 - \xi) \\ \xi \end{bmatrix} [(1 - \xi) \; \xi] \, d\xi$$

$$= \begin{bmatrix} m_{11} & m_{12} \\ m_{21} & m_{22} \end{bmatrix} \quad (4.20)$$

where, A = cross-sectional area
ℓ = elemental length
ρ = density

$$m_{11} = \rho A \ell \int_0^1 (1 - \xi)^2 \, d\xi$$

$$= \rho A \ell \int_0^1 (1 + \xi^2 - 2\xi) \, d\xi$$

$$= \rho A \ell \left[\xi + \frac{\xi^3}{3} - \frac{2\xi^2}{2} \right]_0^1$$

$$= \rho A \ell \left(1 + \frac{1}{3} - 1 \right)$$

$$\text{or } m_{11} = \rho A \ell / 3 \quad (4.21)$$

Similarly,

$$m_{12} = m_{21} = \rho A \ell \int_0^1 (1 - \xi) \xi \, d\xi$$

$$= \rho A \ell \int_0^1 (\xi - \xi^2) \, d\xi$$

$$= \rho A \ell \left[\frac{\xi^2}{2} - \frac{\xi^3}{3} \right]_0^1$$

$$= \rho A \ell \left(\frac{1}{2} - \frac{1}{3} \right)$$

$$\text{or } m_{12} = m_{21} = \rho A \ell / 6 \quad (4.22)$$

$$\text{Now } m_{22} = \rho A \ell \int_0^1 \xi^2 \, d\xi$$

$$= \rho A \ell \left[\frac{\xi^3}{3} \right]$$

$$= \rho A \ell \left[\frac{1}{3} \right]$$

Mass Matrix for a Rod Element

$$\text{or } m_{22} = \rho A \ell / 3 \tag{4.23}$$

Substituting equations (4.21) to (4.23) into equation (4.20)

[m] = elemental mass matrix for a rod in one dimension

$$= \frac{\rho AL}{6} \begin{bmatrix} 2 & 1 \\ 1 & 2 \end{bmatrix} \tag{4.24}$$

Similarly, for motion in both the x and y directions, the elemental mass matrix for a **rod in two dimensions**, is given by

$$[m] = \frac{\rho A \ell}{6} \begin{bmatrix} 2 & 0 & 1 & 0 \\ 0 & 2 & 0 & 1 \\ 1 & 0 & 2 & 0 \\ 0 & 1 & 0 & 2 \end{bmatrix} \begin{matrix} u_i \\ v_i \\ u_j \\ v_j \end{matrix} \tag{4.25}$$

with column labels $u_i \; v_i \; u_j \; v_j$

and for motion in the x, y and z directions, the elemental mass matrix for a **rod in three dimensions**, is given by

$$[m] = \frac{\rho A \ell}{6} \begin{bmatrix} 2 & 0 & 0 & 1 & 0 & 0 \\ 0 & 2 & 0 & 0 & 1 & 0 \\ 0 & 0 & 2 & 0 & 0 & 1 \\ 1 & 0 & 0 & 2 & 0 & 0 \\ 0 & 1 & 0 & 0 & 2 & 0 \\ 0 & 0 & 1 & 0 & 0 & 2 \end{bmatrix} \begin{matrix} u_i \\ v_i \\ w_i \\ u_j \\ v_j \\ w_j \end{matrix} \tag{4.26}$$

with column labels $u_i \; v_i \; w_i \; u_j \; v_j \; w_j$

4.3.3 To obtain $[m^\circ]$

From equation (2.37) in terms of the **global co-ordinates**, the elemental mass matrix for a **rod in two dimensions** is given by

$$[m^\circ] = \begin{bmatrix} T & 0_2 \\ 0_2 & T \end{bmatrix}^T [m] \begin{bmatrix} T & 0_2 \\ 0_2 & T \end{bmatrix} \qquad (4.27)$$

where from equations (2.30)

$$[T] = \begin{bmatrix} c & s \\ -s & c \end{bmatrix} \qquad (4.28)$$

$c = \cos \alpha$

$s = \sin \alpha$

α = angle of inclination of the rod from the horizontal axis, positive, counter-clockwise.

Substituting equations (4.25) and (4.29) into equation (4.27), it can be shown that the mass matrix for a **rod element in two dimensions, in global co-ordinates**, $[m^\circ]$, is given by:

$$[m^\circ] = \frac{\rho A \ell}{6} \begin{matrix} & \begin{matrix} u_i^\circ & v_i^\circ & u_j^\circ & v_j^\circ \end{matrix} & \\ & \begin{bmatrix} 2 & 0 & 1 & 0 \\ 0 & 2 & 0 & 1 \\ 1 & 0 & 2 & 0 \\ 0 & 1 & 0 & 2 \end{bmatrix} & \begin{matrix} u_i^\circ \\ v_i^\circ \\ u_j^\circ \\ v_j^\circ \end{matrix} \end{matrix} \qquad (4.29)$$

By a similar process, it can be shown that the mass matrix for a **rod element in three dimensions, in global co-ordinates** $[m^\circ]$, is given by:

$$[m^o] = \frac{\rho A \ell}{6} \begin{bmatrix} 2 & 0 & 0 & 1 & 0 & 0 \\ 0 & 2 & 0 & 0 & 1 & 0 \\ 0 & 0 & 2 & 0 & 0 & 1 \\ 1 & 0 & 0 & 2 & 0 & 0 \\ 0 & 1 & 0 & 0 & 2 & 0 \\ 0 & 0 & 1 & 0 & 0 & 2 \end{bmatrix} \begin{matrix} u_i^o \\ v_i^o \\ w_i^o \\ u_j^o \\ v_j^o \\ w_j^o \end{matrix} \quad (4.30)$$

(columns labelled $u_i^o\ v_i^o\ w_i^o\ u_j^o\ v_j^o\ w_j^o$)

4.3.4 Added Masses

The mass matrices of equations (4.29) and (4.30) only allow for the self mass of the structure. In cases where there is an additional concentrated mass, namely M_i, attached to node i, the effects of this mass must be added to the structure at node i in the following manner.

$$\begin{bmatrix} M_i & 0 \\ 0 & M_i \end{bmatrix} \begin{matrix} u_i^o \\ v_i^o \end{matrix} \quad \text{in two dimensions} \quad (4.31)$$

and

$$\begin{bmatrix} M_i & 0 & 0 \\ 0 & M_i & 0 \\ 0 & 0 & M_i \end{bmatrix} \begin{matrix} u_i^o \\ v_i^o \\ w_i^o \end{matrix} \quad \text{in three dimensions} \quad (4.32)$$

4.4 Vibrations of Pin-Jointed Trusses

The following worked examples will demonstrate how to apply the theory of section (4.3) to some pin-jointed trusses.

Example 4.1

Determine the natural frequencies and modes of vibration for the rod of Figure 4.2, which has stepped variation in its cross-section. The following may be assumed:

$$E = 1 \times 10^{11} \frac{N}{m^2}$$

$$\rho = \text{density} = \frac{7860 \text{kg}}{m^3}$$

$$A_{1-2} = \text{cross-sectional area of element } 1-2$$
$$= 0.001 m^2$$

$$A_{2-3} = \text{cross-sectional area of element } 2-3$$
$$= 0.0006 m^2$$

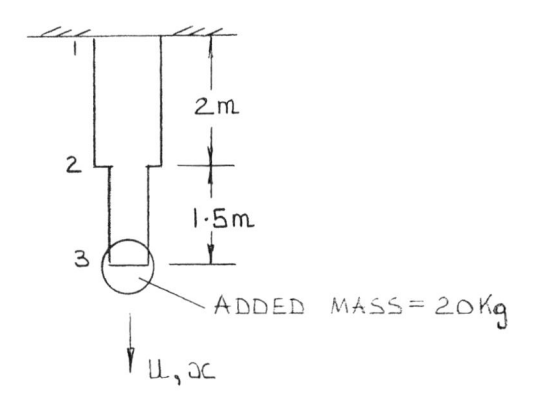

Figure 4.2 Rod structure.

Element 1-2

From equation (2.9), the elemental stiffness matrix for element 1 - 2 =

$$[k_{1-2}] = \frac{0.001 \times 2E11}{2} \begin{bmatrix} 1 \end{bmatrix} \begin{matrix} u_2 \\ u_2 \end{matrix}$$

Sec. 4.4]   Vibrations of Pin-Jointed Trusses   117

$$= \begin{bmatrix} u_2 \\ 100\ E6 \end{bmatrix} u_2 \qquad (4.33)$$

Similarly, from equation (4.24), the elemental mass matrix for element 1 - 2 =

$$[m_{1-2}] = \frac{7860 \times 0.001 \times 2}{6} \begin{bmatrix} u_2 \\ 2 \end{bmatrix} u_2$$

$$= \begin{bmatrix} u_2 \\ 5.24 \end{bmatrix} u_2 \qquad (4.34)$$

Element 2-3

From equation (2.9), the elemental stiffness matrix for element 2 - 3 =

$$[k_{2-3}] = \frac{0.0006 \times 2E11}{1.5} \begin{bmatrix} u_2 & u_3 \\ 1 & -1 \\ -1 & 1 \end{bmatrix} \begin{matrix} u_2 \\ u_3 \end{matrix}$$

$$= \begin{bmatrix} u_2 & u_3 \\ 80E6 & -80E6 \\ -80E6 & 80E6 \end{bmatrix} \begin{matrix} u_2 \\ u_3 \end{matrix} \qquad (4.35)$$

From equation 4.24, the elemental mass matrix for element 2 - 3 =

$$[m_{2-3}] = \frac{7860 \times 0.0006 \times 1.5}{6} \begin{bmatrix} u_2 & u_3 \\ 2 & 1 \\ 1 & 2 \end{bmatrix} \begin{matrix} u_2 \\ u_3 \end{matrix}$$

$$= \begin{bmatrix} u_2 & u_3 \\ 2.358 & 1.179 \\ 1.179 & 2.358 \end{bmatrix} \begin{matrix} u_2 \\ u_3 \end{matrix} \qquad (4.36)$$

The system stiffness matrix $[K_{11}]$, corresponding to the free displacements, namely u_2 and u_3, can be obtained by adding together the stiffness influence coefficients, corresponding to these displacements, from equations (4.33) and (4.35), as shown below.

$$[K_{11}] = \begin{bmatrix} 100\,E6 & -80\,E6 \\ +80\,E6 & \\ \hline -80\,E6 & 80\,E6 \end{bmatrix} \begin{matrix} u_2 \\ \\ u_3 \end{matrix}$$

$$= 1E6 \begin{bmatrix} 180 & -80 \\ -80 & 80 \end{bmatrix} \begin{matrix} u_2 \\ u_3 \end{matrix} \tag{4.37}$$

Similarly, the system mass matrix $[M_{11}]$ corresponding to the free displacement, namely u_2 and u_3, can be obtained by adding together the coefficients of mass, corresponding to these displacements, from equations (4.34) and (4.36), together with the additional concentrated mass at node 3, as shown below.

$$[M_{11}] = \begin{bmatrix} 5.24 & 1.179 \\ +2.358 & \\ \hline 1.179 & 2.358 \end{bmatrix} \begin{matrix} u_2 \\ \\ u_3 \end{matrix} + \begin{bmatrix} 0 & 0 \\ \hline 0 & 20 \end{bmatrix} \begin{matrix} u_2 \\ \\ u_3 \end{matrix}$$

$$= \begin{bmatrix} 7.598 & 1.179 \\ 1.179 & 22.358 \end{bmatrix} \begin{matrix} u_2 \\ u_3 \end{matrix} \tag{4.38}$$

Now from Newton's 2nd Law of motion,

$$m\frac{d^2u}{dt^2} + ku = 0 \tag{4.39}$$

but $u = ce^{j\omega t}$

and $\dfrac{d^2u}{dt^2} = -\omega^2 ce^{j\omega t} = -\omega^2 u$ (4.40)

Substituting equation (4.40) into equation (4.39)

$-\omega^2 mu + ku = 0$

or in matrix form

$$[K]\{U_i\} - \omega^2 [M]\{U_i\} = \{0\}$$

The condition $\{U_i\} = \{0\}$ is not of practical interest

$$\therefore \ |[K] - \omega^2 [M]| = 0 \quad (4.41)$$

or for **constrained** or **over-constrained** structures

$$|K_{11} - \omega^2 [M_{11}]| = 0 \quad (4.42)$$

Substituting equations (4.37) and (4.38) into equation (4.42)

$$\left| \begin{bmatrix} 180E6 & -80E6 \\ -80E6 & 80E6 \end{bmatrix} - \omega^2 \begin{bmatrix} 7.598 & 1.179 \\ 1.179 & 22.358 \end{bmatrix} \right| = 0 \quad (4.43)$$

or

$$\left| \begin{matrix} (180E6 - 7.598\omega^2) & -(80E6 + 1.179\omega^2) \\ -(80E6 + 1.179\omega^2) & (80E6 - 22.358\omega^2) \end{matrix} \right| = 0$$

Expanding the above determinant,

$(180\ E6 - 7.598\ \omega^2)(80\ E6 - 22.358\ \omega^2)$

$- (80\ E6 + 1.179\ \omega^2)^2 = 0$

$1.44\ E16 - 4.632\ E9\ \omega^2 + 169.9\ \omega^4$

$- 6.4\ E15 - 188.64\ E6\ \omega^2 - 1.39\ \omega^4 = 0$

or $168.51\ \omega^4 - 4.821\ E9\ \omega^2 + 8\ E15 = 0$ (4.44)

Solving the equation (4.44)

$$\omega^2 = \frac{4.821E9 \pm \sqrt{(4.821E9)^2 - 4 \times 8E15 \times 168.51}}{2 \times 168.51}$$

$$= \frac{4.821E9 \pm 4.225E9}{337.02}$$

Hence,

$\omega_1 = 1329.8$ rads/s and $\omega_2 = 5180.8$ rads/s

$n = \omega/2\pi$

$\therefore n_1 = 211.6$ Hz

and $n_2 = 824.6$ Hz

To determine eigenmodes

To determine the first eigenmode, substitute ω_1^2 into the first line of equation (4.43)

$(180\ E6 - 7.598 \times 1.768\ E6)\ u_2$

$- (80\ E6 + 1.179 \times 1.768\ E6)\ u_3 = 0$

$166.6\ u_2 - 82.1\ u_3 = 0$

Let $u_3 = 1$ $\therefore u_2 = 0.492$

Hence, the **first eigenmode** is

$[u_2 \quad u_3] = [0.492 \quad 1]$

To determine the **second eigenmode**, substitute ω_2^2 into the second line of equation (4.43)

$- (80\ E6 + 1.179 \times 26.84\ E6)\ u_2$

$+ (80\ E6 - 22.358 \times 26.84\ E6)\ u_3 = 0$

or

$- 111.6\ u_2 - 520.1\ u_3 = 0$

Let $u_2 = 1$ $\therefore u_2 = -0.215$

$\therefore$ the **second eigenmode** is

$[u_2 \quad u_3] = [1 \quad -0.215]$

Sec. 4.4] Vibrations of Pin-Jointed Trusses 121

NB It should be noted that in this case, the two eigenmodes could have been obtained by substituting ω_1^2 into the second line of equation (4.43), and by substituting ω_2^2 into the first line of equation (4.43).

Example 4.2

Determine the natural frequencies of vibration and eigenmodes for the plane pin-jointed truss of Figure 4.3. It may be assumed that for all members

$$A = 0.0006 \text{ m}^2 \qquad E = 2 \times 10^{11} \text{ N/m}^2 \qquad \rho = 7860 \text{ kg/m}^3$$

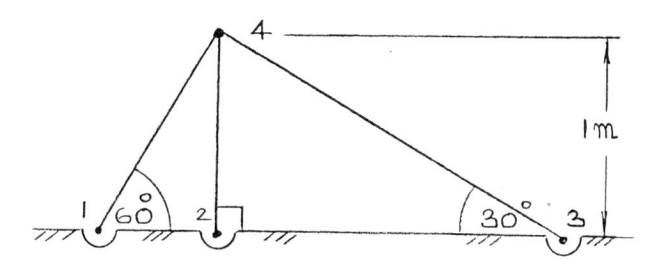

Figure 4.3 Plane pin-jointed truss.

Element 1-4

$\alpha = 60°$ $\qquad$ $c = 0.5$ $\qquad$ $s = 0.866$ $\qquad$ $\ell_{1-4} = 1/\sin 60 = 1.155$ m

Substituting the above into equations (2.43) and (4.29)

$$[k_{1-4}^o] = \frac{0.0006 \times 2E11}{1.155} \begin{bmatrix} 0.25 & 0.433 \\ 0.433 & 0.75 \end{bmatrix}$$

$$= \begin{bmatrix} 2.597E7 & 4.499E7 \\ 4.499E7 & 7.792E7 \end{bmatrix} \begin{matrix} u_4^o \\ c_4^o \end{matrix} \qquad (4.45)$$

$$\text{and } [m_{1-4}^o] = \frac{7860 \times 0.0006 \times 1.155}{6} \begin{bmatrix} 2 & 0 \\ 0 & 2 \end{bmatrix}$$

$$\begin{bmatrix} \overset{o}{u_4} & \overset{o}{v_4} \\ 1.816 & 0 \\ 0 & 1.816 \end{bmatrix} \begin{matrix} \overset{o}{u_4} \\ \overset{o}{v_4} \end{matrix} \qquad (4.46)$$

Element 2-4

$\alpha = 90°$ $\qquad c = 0 \qquad s = 1 \qquad \ell_{2\text{-}4} = 1m$

Substituting the into equations (2.43) and (4.29)

$$\left[k_{2\text{-}4}^o\right] = \frac{0.0006 \times 2E11}{1} \begin{bmatrix} 0 & 0 \\ 0 & 1 \end{bmatrix}$$

$$= \begin{bmatrix} \overset{o}{u_4} & \overset{o}{v_4} \\ 0 & 0 \\ 0 & 1.2E8 \end{bmatrix} \begin{matrix} \overset{o}{u_4} \\ \overset{o}{v_4} \end{matrix} \qquad (4.47)$$

$$\left[m_{2\text{-}4}^o\right] = \frac{7860 \times 0.0006 \times 1}{6} \begin{bmatrix} 2 & 0 \\ 0 & 2 \end{bmatrix}$$

$$= \begin{bmatrix} \overset{o}{u_4} & \overset{o}{v_4} \\ 1.572 & 0 \\ 0 & 1.572 \end{bmatrix} \begin{matrix} \overset{o}{u_4} \\ \overset{o}{v_4} \end{matrix} \qquad (4.48)$$

Element 3-4

$\alpha = 150°$ $\qquad c = -0.866 \qquad s = 0.5 \qquad \ell_{3\text{-}4} = 2m$

From equations (2.43) and (4.30)

$$= \begin{bmatrix} \overset{u_4^o}{4.5\text{E}7} & \overset{v_4^o}{-2.598\text{E}7} \\ -2.598\text{E}7 & 1.5\text{E}7 \end{bmatrix} \begin{matrix} u_4^o \\ v_4^o \end{matrix} \quad (4.49)$$

$$\left[m_{3-4}^o \right] = \frac{7860 \times 0.0006 \times 2}{6} \begin{bmatrix} 2 & 0 \\ 0 & 2 \end{bmatrix}$$

$$= \begin{bmatrix} \overset{u_4^o}{3.144} & \overset{v_4^o}{0} \\ 0 & 3.144 \end{bmatrix} \begin{matrix} u_4^o \\ v_4^o \end{matrix} \quad (4.50)$$

From equations (4.45), (4.47) and (4.49)

$$[K_{11}] = \begin{bmatrix} \begin{array}{c} 2.597\ \text{E}7 \\ +o \\ +4.5\ \text{E}7 \end{array} & \begin{array}{c} 4.499\ \text{E}7 \\ +0 \\ -2.598\ \text{E}7 \end{array} \\ \begin{array}{c} 4.499\ \text{E}7 \\ +o \\ -2.598\ \text{E}7 \end{array} & \begin{array}{c} 7.792\ \text{E}7 \\ +1.2\ \text{E}8 \\ +1.5\ \text{E}7 \end{array} \end{bmatrix} \begin{matrix} u_4^o \\ \\ v_4^o \end{matrix} \quad (4.51)$$

$$= \begin{bmatrix} \overset{u_4^o}{7.097\text{E}7} & \overset{v_4^o}{1.901\text{E}7} \\ 1.901\text{E}7 & 2.129\text{E}8 \end{bmatrix} \begin{matrix} u_4^o \\ v_4^o \end{matrix} \quad (4.52)$$

From equations (4.46), (4.48) and (4.50)

$$[M_{11}] = \begin{bmatrix} \begin{array}{c|c} \begin{matrix} 1.816 \\ +1.572 \\ +3.144 \end{matrix} & \begin{matrix} 0 \\ +0 \\ +0 \end{matrix} \\ \hline \begin{matrix} 0 \\ +0 \\ +0 \end{matrix} & \begin{matrix} 1.816 \\ +1.572 \\ 3.144 \end{matrix} \end{array} \end{bmatrix} \begin{matrix} u_4^o \\ \\ v_4^o \end{matrix}$$

$$\begin{matrix} u_4^o & v_4^o \end{matrix} \qquad (4.53)$$

$$[M_{11}] = \begin{bmatrix} 6.532 & 0 \\ 0 & 6.532 \end{bmatrix} \begin{matrix} u_4^o \\ v_4^o \end{matrix} \qquad (4.54)$$

Now $|[K_{11}] - \omega^2 [M_{11}]| = 0$

or

$$\left| \begin{bmatrix} 7.097\text{E}7 & 1.901\text{E}7 \\ 1.901\text{E}7 & 2.129\text{E}8 \end{bmatrix} - \omega^2 \begin{bmatrix} 6.532 & 0 \\ 0 & 6.532 \end{bmatrix} \right| = 0$$

or

$$\begin{vmatrix} (7.097\text{E}7 - 6.532\omega^2) & (1.901\text{E}7) \\ (1.901\text{E}7) & (2.129\text{E}8 - 6.532\omega^2) \end{vmatrix} = 0 \qquad (4.55)$$

Expanding the determinant of equation (4.55),

$(7.097 \text{ E}7 - 6.532 \, \omega^2)(2.129 \text{ E}8 - 6.532 \, \omega^2)$

$-(1.901 \text{ E}7)^2 = 0$

or $1.511 \text{ E}16 - 1.854 \text{ E}9 \, \omega^2 + 42.667 \, \omega^4$

$- 3.614 \text{ E}14 = 0$

$42.667 \, \omega^4 - 1.854 \text{ E}9 \, \omega^2 + 1.475 \text{ E}16 = 0$

Sec. 4.4] Vibrations of Pin-Jointed Trusses 125

Solving the above quadratic equation,

$$\omega^2 = \frac{1.854\text{E}9 \pm \sqrt{(3.437\text{E}18 - 2.517\text{E}18)}}{85.334}$$

$$\omega^2 = \frac{1.854\text{E}9 \pm 9.59\text{E}8}{85.334}$$

$\omega_1^2 = 1.0488\text{ E}7$; $\omega_1 = 3238.5$ rads/s; $n_1 = 515.4$ Hz

$\omega_2^2 = 3.296\text{ E}7$; $\omega_2 = 5741.5$ rads/s; $n_2 = 913.8$ Hz

1st Eigenmode

Substituting ω_1^2 into the first line of equation (4.55)

$$2.462\text{ E}6\ u_4^\circ + 1.901\text{ E}7\ v_4^\circ = 0$$

Let $u_4^\circ = 1$ $\therefore v_4^\circ = -0.13$

$$\therefore [u_4^\circ\ v_4^\circ] = [1\ \ -0.13]$$

2nd Eigenmode

Substituting ω_2^2 into the second line of equation (4.55)

$$1.901\text{ E}7\ u_4^\circ - 2.395\text{ E}6\ v_4^\circ = 0$$

Let $v_4^\circ = 1$ $\therefore u_4^\circ = 0.126$

$$\therefore [u_4^\circ\ v_4^\circ] = [0.126\ \ 1]$$

These two eigenmodes are shown diagrammatically in Figures 4.4 and 4.5.

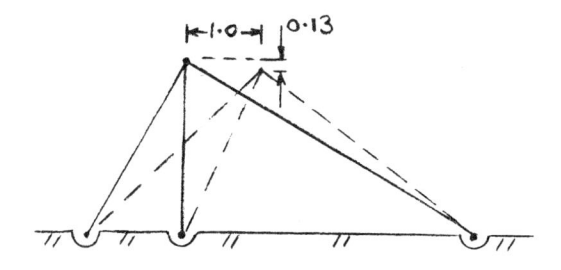

Figure 4.4 First Eigenmode.

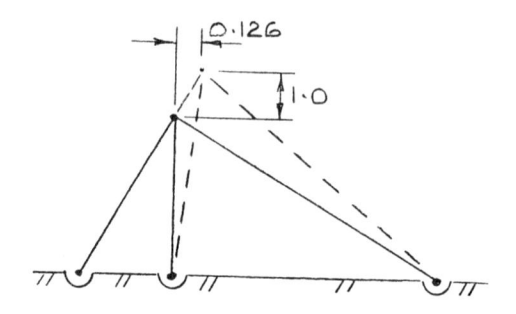

Figure 4.5 Second Eigenmode.

Example 4.3

Determine the natural frequencies of vibration of the plane pin-jointed truss of Figure 4.3, if a mass of 10 kg is added to node 4.

From equation (4.52)

$$[K_{11}] = \begin{matrix} u_4^o & v_4^o \\ \begin{bmatrix} 7.097E7 & 1.901E7 \\ 1.901E7 & 2.129E8 \end{bmatrix} & \begin{matrix} u_4^o \\ u_4^o \end{matrix} \end{matrix} \qquad (4.56)$$

From equations (4.53) and (4.31), the system mass matrix, namely $[M_{11}]$, becomes

$$[M_{11}] = \begin{matrix} u_4^o & v_4^o \\ \begin{bmatrix} 6.532 + 10 & 0 \\ 0 & 6.532 + 10 \end{bmatrix} & \begin{matrix} u_4^o \\ v_4^o \end{matrix} \end{matrix} \qquad (4.57)$$

$$= \begin{bmatrix} 16.532 & 0 \\ 0 & 16.532 \end{bmatrix} \qquad (4.58)$$

Now $|\,[K_{11}] - \omega^2 [M_{11}]\,| = 0$

$$\left|\begin{bmatrix} 7.097\text{E}7 & 1.901\text{E}7 \\ 1.901\text{E}7 & 2.129\text{E}8 \end{bmatrix} - \omega^2 \begin{bmatrix} 16.532 & 0 \\ 0 & 16.532 \end{bmatrix}\right| = 0$$

$$\left|\begin{matrix} (7.097\text{E}7 - 16.532\omega^2) & 1.901\text{E}7 \\ 1.901\text{E}7 & (2.129\text{E}8 - 16.532\omega^2) \end{matrix}\right| = 0 \qquad (4.59)$$

or $(7.097\,\text{E}7 - 16.532\,\omega^2)(2.129\,\text{E}8 - 16.532\,\omega^2)$

$- (1.901\,\text{E}7)^2 = 0$

$1.511\,\text{E}16 + 273.3\,\omega^4 - 4.693\,\text{E}9\,\omega^2 - 3.613\,\text{E}14 = 0$

$273.3\,\omega^4 - 4.693\,\text{E}9\,\omega^2 + 1.475\,\text{E}16 = 0$

$\therefore \omega^2 = \dfrac{4.693\,\text{E}9 \pm \sqrt{(2.202\,\text{E}19 - 1.612\,\text{E}19)}}{546.6}$

or $\omega^2 = \dfrac{4.693\,\text{E}9 \pm 2.429\,\text{E}9}{546.6}$

Hence,

$\omega_1^2 = 4.142\,\text{E}6;\qquad \omega_1 = 2035\text{ rads/s};\qquad n_1 = 323.9\text{ Hz}$

and $\omega_2^2 = 1.303\,\text{E}7;\qquad \omega_2 = 3610\text{ rads/s};\qquad n_2 = 574.5\text{ Hz}$

Example 4.4

Determine the natural frequencies of vibration and the eigenmodes of the pin-jointed space truss of Figure 4.6. It may be assumed that:

$A = 0.001\text{ m}^2,\qquad E = 2 \times 10^{11}\text{ N/m}^2\qquad \text{and}\quad \rho = 7860\text{ kg/m}^3$

128   **Vibration of Structures**   Ch.4

(a) Plan

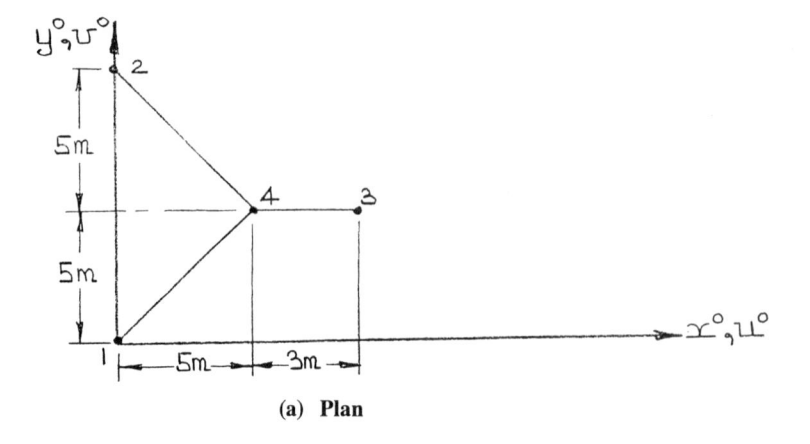

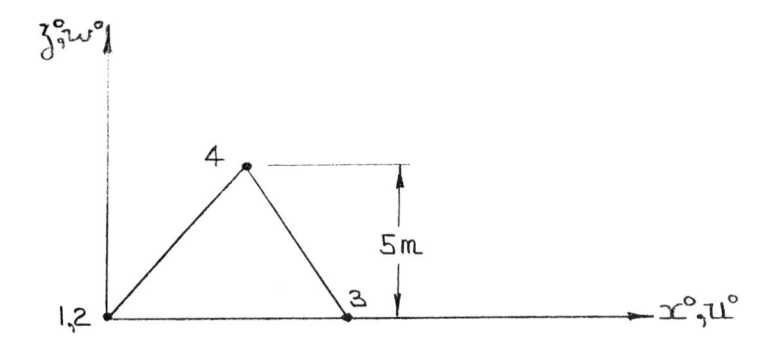

(b) Front View

Figure 4.6 Pin-jointed tripod.

This is a four dimensional problem, but its solution is similar to that used for Examples 4.2 and 4.3.

Element 1-4

$x_1^o = y_1^o = z_1^o = 0;$ $x_4^o = 5m;$ $y_4^o = 5m;$ $z_4^o = 5m$

From equation (2.66)

$$\ell_{1-4} = \sqrt{\left[\left(x_4^o - x_1^o\right)^2 + \left(y_4^o - y_1^o\right)^2 + \left(z_4^o - z_1^o\right)^2\right]}$$

$$= \sqrt{(5^2 + 5^2 + 5^2)} = 8.66m$$

Sec. 4.4] Vibrations of Pin-Jointed Trusses

From equation (2.65)

$$C_{x,x^o} = \frac{(x_4^o - x_1^o)}{\ell_{1-4}}$$

$$= \frac{5}{8.66} = 0.5774$$

$$C_{x,y^o} = \frac{(y_4^o - y_1^o)}{\ell_{1-4}}$$

$$= \frac{5}{8.66} = 0.5774 \qquad (4.60)$$

$$C_{x,z^o} = \frac{(z_4^o - z_1^o)}{\ell_{1-4}}$$

$$= \frac{5}{8.66} = 0.5774$$

Substituting the above data into equation (2.63)

$$[k_{1-4}^o] = \frac{0.001 \times 2E11}{8.66} \begin{matrix} & u_4^o & v_4^o & w_4^o \\ \begin{bmatrix} 0.333 & 0.333 & 0.333 \\ 0.333 & 0.333 & 0.333 \\ 0.333 & 0.333 & 0.333 \end{bmatrix} & \begin{matrix} u_4^o \\ v_4^o \\ w_4^o \end{matrix} \end{matrix}$$

$$= 1E6 \begin{matrix} & u_4^o & v_4^o & w_4^o \\ \begin{bmatrix} 7.7 & 7.7 & 7.7 \\ 7.7 & 7.7 & 7.7 \\ 7.7 & 7.7 & 7.7 \end{bmatrix} & \begin{matrix} u_4^o \\ v_4^o \\ w_4^o \end{matrix} \end{matrix} \qquad (4.61)$$

From equation (4.30)

$$[m_{1-4}^o] = \frac{7860 \times 0.001 \times 8.66}{6} \begin{bmatrix} 2 & 0 & 0 \\ 0 & 2 & 0 \\ 0 & 0 & 2 \end{bmatrix}$$

$$= \begin{bmatrix} 22.69 & 0 & 0 \\ 0 & 22.69 & 0 \\ 0 & 0 & 22.69 \end{bmatrix} \begin{matrix} u_4^o \\ v_4^o \\ w_4^o \end{matrix} \quad \begin{matrix} u_4^o & u_4^o & w_4^o \end{matrix} \tag{4.62}$$

Element 2-4

$$x_2^o = 0; \quad y_2^o = 10 \text{ m}; \quad z_2^o = 0; \quad x_4^o = y_4^o = z_4^o = 5\text{m}$$

From equation (2.66)

$$\ell_{2-4} = \sqrt{[5^2 + (5-10)^2 + 5^2]} = 8.66\text{m}$$

From equation (2.65)

$$C_{x,x^o} = \frac{5}{8.66} = 0.5774$$

$$C_{x,y^o} = \frac{(5-10)}{8.66} = -0.5774$$

$$C_{x,z^o} = \frac{5}{8.66} = 0.5774$$

Substituting the above data into equation (2.63)

$$[k_{2-4}^o] = \frac{0.001 \times 2E11}{8.66} \begin{bmatrix} u_4^o & v_4^o & w_4^o \\ 0.333 & -0.333 & 0.333 \\ -0.333 & 0.333 & -0.333 \\ 0.333 & -0.333 & 0.333 \end{bmatrix} \begin{matrix} u_4^o \\ v_4^o \\ w_4^o \end{matrix}$$

$$[k_{2-4}^o] = 1E6 \begin{bmatrix} u_4^o & v_4^o & w_4^o \\ 7.7 & -7.7 & 7.7 \\ -7.7 & 7.7 & -7.7 \\ 7.7 & -7.7 & 7.7 \end{bmatrix} \begin{matrix} u_4^o \\ v_4^o \\ w_4^o \end{matrix} \quad (4.63)$$

From equation (4.30)

$$[m_{2-4}^o] = \frac{7860 \times 0.001 \times 8.66}{6} \begin{bmatrix} 2 & 0 & 0 \\ 0 & 2 & 0 \\ 0 & 0 & 2 \end{bmatrix}$$

$$= \begin{bmatrix} u_4^o & v_4^o & w_4^o \\ 22.69 & 0 & 0 \\ 0 & 22.69 & 0 \\ 0 & 0 & 22.69 \end{bmatrix} \begin{matrix} u_4^o \\ v_4^o \\ w_4^o \end{matrix} \quad (4.64)$$

Element 3-4

$x_3^o = 8m$; $y_3^o = 5m$; $z_3^o = 0$; $x_4^o = 5m$; $y_4^o = 5m$; $z_4^o = 5m$

From equation (2.66)

$$\ell_{3-4} = \sqrt{(5-8)^2 + (5-5)^2 + (5-0)^2} = 5.831m$$

From equation (2.65)

$$C_{x,x^o} = \frac{(5-8)}{5.831} = -0.514$$

$$C_{x,y^o} = \frac{0}{5.831} = 0$$

$$C_{x,y^o} = \frac{(5-0)}{5.831} = 0.857$$

Substituting the above data into equation (2.63)

$$\left[k_{3-4}^o\right] = \frac{0.001 \times 2E11}{5.831} \begin{bmatrix} 0.264 & 0 & -0.440 \\ 0 & 0 & 0 \\ -0.440 & 0 & 0.734 \end{bmatrix} \begin{matrix} u_4^o \\ v_4^o \\ w_4^o \end{matrix}$$

$$= 1E6 \begin{bmatrix} 9.055 & 0 & -15.092 \\ 0 & 0 & 0 \\ -15.092 & 0 & 25.176 \end{bmatrix} \begin{matrix} u_4^o \\ v_4^o \\ w_4^o \end{matrix} \quad (4.65)$$

Sec. 4.4] Vibrations of Pin-Jointed Trusses

Substituting the above into equation (4.30)

$$[m^o_{3-4}] = \frac{7860 \times 0.001 \times 5.831}{6} \begin{bmatrix} 2 & 0 & 0 \\ 0 & 2 & 0 \\ 0 & 0 & 2 \end{bmatrix}$$

$$= \begin{bmatrix} 15.277 & 0 & 0 \\ 0 & 15.277 & 0 \\ 0 & 0 & 15.277 \end{bmatrix} \begin{matrix} u^o_4 \\ v^o_4 \\ w^o_4 \end{matrix} \qquad (4.66)$$

with columns $u^o_4 \; v^o_4 \; w^o_4$

From equations (4.61), (4.63) and (4.65)

$$[K_{11}] = 1E6 \begin{bmatrix} 7.7+7.7 & 7.7-7.7 & 7.7+7.7 \\ +9.055 & +0 & -15.092 \\ \hline 7.7-7.7 & 7.7+7.7 & 7.7-7.7 \\ +0 & +0 & +0 \\ \hline 7.7+7.7 & 7.7-7.7 & 7.7+7.7 \\ -15.092 & +0 & +25.176 \end{bmatrix} \begin{matrix} u^o_4 \\ v^o_4 \\ w^o_4 \end{matrix} \qquad (4.67)$$

with columns $u^o_4 \; v^o_4 \; w^o_4$

$$= 1E6 \begin{bmatrix} 24.45 & 0 & 0.308 \\ 0 & 15.4 & 0 \\ 0.308 & 0 & 40.58 \end{bmatrix} \begin{matrix} u^o_4 \\ v^o_4 \\ w^o_4 \end{matrix} \qquad (4.68)$$

From equations (4.62), (4.64) and (4.66)

	u_4^o	v_4^o	w_4^o	
$[M_{11}] =$	22.69 + 22.69 15.277	0	0	u_4^o
	0	22.69 + 22.69 15.277	0	v_4^o
	0	0	22.69 + 22.69 +15.277	w_4^o

$$= \begin{bmatrix} 60.66 & 0 & 0 \\ 0 & 60.66 & 0 \\ 0 & 0 & 60.66 \end{bmatrix} \qquad (4.69)$$

Now $\left| [K_{11}] - \omega^2 [M_{11}] \right| = 0$

$$\text{or} \quad \left| 1E6 \begin{bmatrix} 24.45 & 0 & 0.308 \\ 0 & 15.4 & 0 \\ 0.308 & 0 & 40.58 \end{bmatrix} - \omega^2 \begin{bmatrix} 60.66 & 0 & 0 \\ 0 & 60.66 & 0 \\ 0 & 0 & 60.66 \end{bmatrix} \right| \qquad (4.70)$$

Expanding out the 2nd line of equation (4.70)

$$15.4\ E6 - 60.66\ \omega^2 = 0$$

$$\therefore \omega_1^2 = 2.539\ E5; \quad \omega_1 = 503.9 \text{ rads/s}; \quad n_1 = 80.2 \text{ Hz}$$

Hence, equation (4.70) becomes

$$\left| 1E6 \begin{bmatrix} 24.45 & 0.308 \\ 0.308 & 40.58 \end{bmatrix} - \omega^2 \begin{bmatrix} 60.66 & 0 \\ 0 & 60.66 \end{bmatrix} \right| = 0$$

or
$$\begin{vmatrix} (24.45E6 - 60.66\,\omega^2) & 0.308E6 \\ 0.308E6 & (40.58E6 - 60.66\,\omega^2) \end{vmatrix} = 0 \quad (4.71)$$

or $(24.45E6 - 60.66\,\omega^2)(40.58E6 - 60.66\,\omega^2) - (0.308E6)^2 = 0$

$9.922E14 + 3680\,\omega^4 - 3.945E9\,\omega^2 - 9.486E10 = 0$

or $3680\,\omega^4 - 3.945E9\,\omega^2 + 9.921E14 = 0$

$$\omega^2 = \frac{3.945E9 \pm \sqrt{1.556E19 - 1.460E19}}{7360}$$

$$= \frac{3.945E9 \pm 9.779E8}{7360}$$

$\omega^2 = 4.031E5$ or $6.669E5$

$\omega_2 = 634.9$ rads/s; $n_2 = 101$Hz

$\omega_3 = 816.6$ rads/s; $n_3 = 130$Hz

Eigenmodes

<u>1st</u> By inspection

$$[u_4^\circ \quad v_4^\circ \quad w_4^\circ] = [0 \quad 1 \quad 0]$$

<u>2nd</u> Substitute ω_2^2 into the first line of equation (4.71)

$$-2046\,u_4^\circ + 0.308E6\,w_4^\circ = 0$$

Let $u_4^\circ = 1$ $\therefore w_4^\circ = 0.007$

$\therefore [u_4^\circ \quad v_4^\circ \quad w_4^0] = [1 \quad 0 \quad 0.007]$

3rd Substitute ω_3^2 into the second line of equation (4.71)

$$0.308E6\, u_4^o + 1.26E5\, w_4^o = 0$$

Let $w_4^o = 1$ ∴ $u_4^o = -0.41$

∴ $\begin{bmatrix} u_4^o & v_4^o & w_4^o \end{bmatrix} = \begin{bmatrix} -0.41 & 0 & 1.0 \end{bmatrix}$

Example 4.5

Determine the natural frequencies of vibration of the pin-jointed space truss of Example 4.4, if a concentrated mass of 40 kg is added to node 4.

From equation (4.68)

$$[K_{11}] = \begin{matrix} & \begin{matrix} u_4^o & v_4^o & w_4^o \end{matrix} \\ & \begin{bmatrix} 24.45 & 0 & 0.308 \\ 0 & 15.4 & 0 \\ 0.308 & 0 & 40.58 \end{bmatrix} \begin{matrix} u_4^o \\ v_4^o \\ w_4^o \end{matrix} \end{matrix} \quad (4.72)$$

From equations (4.69) and (4.32)

$$[M_{11}] = \begin{bmatrix} 60.66 + 40 & 0 & 0 \\ 0 & 60.66 + 40 & 0 \\ 0 & 0 & 60.66 + 40 \end{bmatrix} \quad (4.73)$$

$$= \begin{matrix} & \begin{matrix} u_4^o & v_4^o & w_4^o \end{matrix} \\ & \begin{bmatrix} 100.66 & 0 & 0 \\ 0 & 100.66 & 0 \\ 0 & 0 & 100.66 \end{bmatrix} \begin{matrix} u_4^o \\ v_4^o \\ w_4^o \end{matrix} \end{matrix} \quad (4.74)$$

Sec. 4.4] Vibrations of Pin-Jointed Trusses 137

Now $| [K_{11}] - \omega^2 [M_{11}] | = 0$

or $\left| 1E6 \begin{bmatrix} 24.45 & 0 & 0.308 \\ 0 & 15.4 & 0 \\ 0.308 & 0 & 40.58 \end{bmatrix} - \omega^2 \begin{bmatrix} 100.7 & 0 & 0 \\ 0 & 100.7 & 0 \\ 0 & 0 & 100.7 \end{bmatrix} \right| = 0$ (4.75)

Expanding out the second line of equation (4.75)

$$15.4 \, E6 - 100.7 \, \omega^2 = 0$$

$\therefore \omega_1^2 = 1.529 \, E5; \quad \omega_1 = 391.1 \text{ rads/s}; \quad n_1 = 62.2 \text{ Hz}$

Equation (4.75) now becomes

$$\left| 1E6 \begin{bmatrix} 24.45 & 0.308 \\ 0.308 & 40.58 \end{bmatrix} - \omega^2 \begin{bmatrix} 100.7 & 0 \\ 0 & 100.7 \end{bmatrix} \right| = 0$$

or $\left| \begin{matrix} (24.45E6 - 100.7 \, \omega^2) & 0.308E6 \\ 0.308E6 & (40.58E6 - 100.7 \, \omega^2) \end{matrix} \right| = 0$ (4.76)

Expanding the determinant of equation (4.76)

$(24.45 \, E6 - 100.7 \, \omega^2)(40.58 \, E6 - 100.7 \, \omega^2)$

$- (0.308 \, E6)^2 = 0$

$9.922 \, E14 + 10140 \, \omega^4 - 6.549 \, E9 \, \omega^2 - 9.486 \, E10 = 0$

$1.014 \, E4 \, \omega^4 - 6.549 \, E9 \, \omega^2 + 9.921 \, E14 = 0$

Solving the above quadratic equation,

$$\omega^2 = \frac{6.549E9 \pm \sqrt{4.289E19 - 4.024E19}}{2.028E4}$$

$$= \frac{6.549E9 \pm 1.629E9}{2.028E4}$$

$\omega_2^2 = 2.427\ E5;\qquad \omega_2 = 492.6\ \text{rads/s};\qquad n_2 = 78.4\ \text{Hz}$

$\omega_3^2 = 4.033\ E5;\qquad \omega_3 = 635.0\ \text{rads/s};\qquad n_3 = 101.1\ \text{Hz}$

4.5 Continuous Beams

The beam element, which has four degrees of freedom, namely, v_i, θ_i, v_j and θ_j, is shown in Figure 4.7.

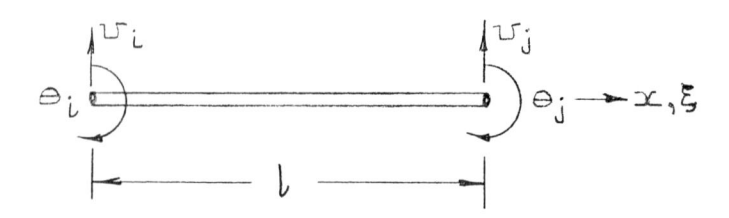

Figure 4.7 Beam element.

4.5.1 To obtain [N]

Now as the beam element of Figure 4.7 has four degrees of freedom, the assumed polynomial for the displacement distortion for "v" must have four constants, as shown by equation (4.77).

$$v = \alpha_1 + \alpha_2 x + \alpha_3 x^2 + \alpha_4 x^3 \qquad (4.77)$$

$$\frac{dv}{dx} = \alpha_2 + 2\alpha_3 x + 3\alpha_4 x^2 \qquad (4.78)$$

The four boundary conditions are:

$$\text{at } x = 0,\ v = v_i \text{ and } \theta_i = -\left(\frac{dv}{dx}\right)_{x=0} \qquad (4.79)$$

$$\text{and at } x = \ell,\ v = v_j \text{ and } \theta_j = -\left(\frac{dv}{dx}\right)_{x=\ell} \qquad (4.80)$$

Sec. 4.5] Continuous Beams 139

It is necessary to have four boundary conditions or boundary values, so that the four constants can be obtained in terms of these four boundary values. Hence, by substituting these four boundary values into equations (4.77) and (4.78), we will obtain four simultaneous equations as follows:

$$v_i = \alpha_1 \tag{4.81}$$

$$\theta_i = -\alpha_2 \tag{4.82}$$

$$v_j = \alpha_1 + \alpha_2 \ell + \alpha_3 \ell^2 + \alpha_4 \alpha^3 \tag{4.83}$$

$$\theta_j = -\alpha_2 - 2\alpha_3 \ell - 3\alpha_4 \ell^2 \tag{4.84}$$

From equations (4.81) and (4.82)

$$\underline{\alpha_1 = v_i} \tag{4.85}$$

and $\underline{\alpha_2 = -\theta_i}$ \hfill (4.86)

Hence, equations (4.83) and (4.84) become

$$v_j = v_i - \theta_i \ell + \alpha_3 \ell^2 + \alpha_4 \ell^3 \tag{4.87}$$

and $\theta_j = +\theta_i - 2\alpha_3 \ell - 3\alpha_4 \ell^2$ \hfill (4.88)

Multiply equation (4.88) by $\ell/2$ to give equation (4.89)

$$\theta_j \frac{\ell}{2} = +\theta_i \frac{\ell}{2} - \alpha_3 \ell^2 - \frac{3\alpha_4 \ell^3}{2} \tag{4.89}$$

Add equation (4.87) to equation (4.89) to eliminate α_3, as follows.

$$v_j + \theta_j \frac{\ell}{2} = v_i - \theta_i \ell + \theta_i \frac{\ell}{2} + \alpha_4 (1 - 3/2) \ell^3$$

or $v_j - v_i + \theta_j \frac{\ell}{2} + \theta_i \frac{\ell}{2} = -\alpha_4 \frac{\ell^3}{2}$

or $\alpha_4 = (-v_j + v_i) \cdot \frac{2}{\ell^3} - \frac{\ell}{2} (\theta_j + \theta_i) \cdot \frac{2}{\ell^3}$

$$\alpha_4 = 2\frac{(v_i - v_j)}{\ell^3} - \frac{(\theta_i + \theta_j)}{\ell^2} \tag{4.90}$$

Substituting equation (4.90) into equation (4.87)

$$v_j = v_i - \theta_i \ell + \alpha_3 \ell^2 + 2(v_i - v_j) - \ell(\theta_i + \theta_j) \tag{4.91}$$

or $v_j - v_i - 2v_i + 2v_j + \theta_i \ell + \theta_i \ell + \theta_j \ell = \alpha_3 \ell^2$

or $\alpha_3 \ell^2 = 3v_j - 3v_i + 2\theta_i \ell + \theta_j \ell$

$$\alpha_3 = 3\frac{(v_j - v_i)}{\ell^2} + \frac{2\theta_i}{\ell} + \frac{\theta_j}{\ell} \tag{4.92}$$

Substituting equations (4.85), (4.86), (4.90) and (4.92) into equation (4.77)

$$v = v_i - \theta_i x + 3(v_j - v_i)\frac{x^2}{\ell^2} + 2\theta_i \frac{x^2}{\ell} + \theta_j x^2/\ell$$

$$+ 2(v_i - v_j)\frac{x^3}{\ell^3} - (\theta_i + \theta_j)\frac{x^3}{\ell^2} \tag{4.93}$$

Let $\xi = x/\ell$

$$\therefore v = v_i - \theta_i \ell\xi + 3(v_j - v_i)\xi^2 + 2\theta_i \ell\xi^2 + \theta_j \ell\xi^2 \tag{4.94}$$

$$+ 2(v_i - v_j)\xi^3 - (\theta_i + \theta_j)\ell\xi^3$$

$$= v_i(1 - 3\xi^2 + 2\xi^3) + \theta_i \ell(-\xi + 2\xi^2 - \xi^3)$$

$$+ v_j(3\xi^2 - 2\xi^3) + \theta_j \ell(\xi^2 - \xi^3)$$

$$= [(1 - 3\xi^2 + 2\xi^3) \quad \ell(-\xi + 2\xi^2 - \xi^3) \quad (3\xi^2 - 2\xi^3) \quad \ell(\xi^2 - \xi^3)] \begin{Bmatrix} v_i \\ \theta_i \\ v_j \\ \theta_j \end{Bmatrix} \tag{4.95}$$

or $u = [N]\{U_i\}$

$\therefore [N]$ = a matrix of shape functions

$$= \left[(1 - 3\xi^2 + 2\xi^3) \quad \ell(-\xi + 2\xi^2 - \xi^3) \quad (3\xi^2 - 2\xi^3) \quad \ell(\xi^2 - \xi^3)\right] \tag{4.96}$$

NB It should be noted that equation (4.95) is a **Hermite polynomial** which osculates to the first degree, ie, it has slope and deflection continuity at the nodes.

4.5.2 To obtain [m]

From Equation (4.11)

$$[m] = \int_{vol} [N]^T \rho \, [N] \, d(vol)$$

$$= \int_o^1 [N]^T \rho \, [N] \, A.\ell.d\xi \tag{4.97}$$

Substituting equation (4.96) into equation (4.97) and integrating between 0 and 1, we get

[m] = mass matrix for a beam element

$$= \frac{\rho A \ell}{420} \begin{array}{c} \begin{matrix} v_i & \theta_i & v_j & \theta_j \end{matrix} \\ \begin{bmatrix} 156 & -22\ell & 54 & 13\ell \\ -22\ell & 4\ell^2 & -13\ell & -3\ell^2 \\ 54 & -13\ell & 156 & 22\ell \\ 13\ell & -3\ell^2 & 22\ell & 4\ell^2 \end{bmatrix} \end{array} \begin{matrix} v_i \\ \theta_i \\ v_j \\ \theta_j \end{matrix} \tag{4.98}$$

Equation (4.98) is the mass matrix of a beam due to itself mass. If, however, a mass of magnitude M_c plus a mass moment of inertia of I_c, were added to node i, it must be added to the mass matrix [m], as follows:

$$\begin{array}{c} \begin{matrix} v_i & \theta_i \end{matrix} \\ \begin{bmatrix} M_c & 0 \\ 0 & I_c \end{bmatrix} \end{array} \begin{matrix} v_i \\ \theta_i \end{matrix} \tag{4.99}$$

where i is the node to which the additional mass is added.

Example 4.6

Determine the natural frequency of vibration for the cantilever of Figure 4.8, which has an additional mass of 5 kg and a mass moment of inertia 0.125 kg m³, added to its free end. It may be assumed that:

$A = 0.001 \text{m}^2 \quad I = 2.1 \times 10^{-7} \text{ m}^4 \quad E = 2 \times 10^{11} \text{ N/m}^2 \quad \rho = 7860 \text{ kg/m}^3$

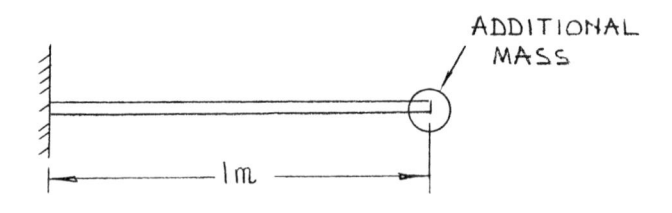

Figure 4.8 Cantilever beam.

From equation (2.90)

$$[K_{11}] = 2 \text{ E}11 \times 2.1 \text{ E}{-7} \begin{array}{c} v_2 \theta_2 \\ \begin{bmatrix} 12 & 6 \\ 6 & 4 \end{bmatrix} \begin{array}{c} v_2 \\ \theta_2 \end{array} \end{array}$$

$$= 1\text{E}6 \begin{array}{c} v_2 \theta_2 \\ \begin{bmatrix} 0.504 & 0.252 \\ 0.252 & 0.168 \end{bmatrix} \begin{array}{c} v_2 \\ \theta_2 \end{array} \end{array} \qquad (4.100)$$

From equation (4.99)

$$[M_{11}] = \frac{7860 \times 1\text{E}{-3} \times 1}{420} \begin{bmatrix} 156 & 22 \\ 22 & 4 \end{bmatrix}$$

Continuous Beams

$$+ \begin{bmatrix} 5 & 0 \\ 0 & 0.125 \end{bmatrix}$$

$$= \begin{bmatrix} 2.919 + 5 & 0.412 \\ 0.412 & 0.075 + 0.125 \end{bmatrix} \quad (4.101)$$

$$= \begin{bmatrix} 7.919 & 0.412 \\ 0.412 & 0.2 \end{bmatrix} \begin{matrix} v_2 \\ \theta_2 \end{matrix} \quad (4.102)$$

Now $|[K_{11}] - \omega^2 [M_{11}]| = 0$

$$\left| 1E6 \begin{bmatrix} 0.504 & 0.252 \\ 0.252 & 0.168 \end{bmatrix} - \omega^2 \begin{bmatrix} 7.919 & 0.412 \\ 0.412 & 0.2 \end{bmatrix} \right| = 0$$

$$\left| \begin{bmatrix} (0.504\ E6 - 7.919\ \omega^2) & (0.252\ E6 - 0.412\ \omega^2) \\ (0.252\ E6 - 0.412\ \omega^2) & (0.168\ E6 - 0.2\ \omega^2) \end{bmatrix} \right| = 0 \quad (4.103)$$

Expanding equation (4.103)

$(0.504\ E6 - 7.919\ \omega^2)(0.168\ E6 - 0.2\ \omega^2)$

$- (0.252\ E6 - 0.412)^2 \omega^2 = 0$

or $8.467\ E10 + 1.584\ \omega^4 - 1.431\ E6\ \omega^2$

$- 6.350\ E10 - 0.17\ \omega^4 + 0.208\ E6\ \omega^2 = 0$

$1.414\ \omega^4 - 1.223\ E6\ \omega^2 + 2.117\ E10 = 0$

Solving the above quadratic equation,

$$\omega^2 = \frac{1.223 \text{ E6} \pm \sqrt{1.496 \text{ E12} - 1.197 \text{ E11}}}{2.828}$$

$$\omega^2 = \frac{1.223 \text{ E6} \pm 1.173 \text{ E6}}{2.828}$$

$\omega_1^2 = 17680$; $\omega_1 = 132.97$ rads/s; $n_1 = 21.16$ Hz

$\omega_2^2 = 847242$; $\omega_2 = 920.45$ rads/s; $n_2 = 146.5$ Hz

Example 4.7

Determine the resonant frequencies of vibration for the continuous beam of Figure 4.9, which is simply supported at nodes 2 and 3, given that:

$A = 0.001 \text{m}^2$; $I = 2\text{E-}7 \text{ m}^4$; $E = 2 \text{ E11 N/m}^2$; $\rho = 7860 \text{ kg/m}^3$

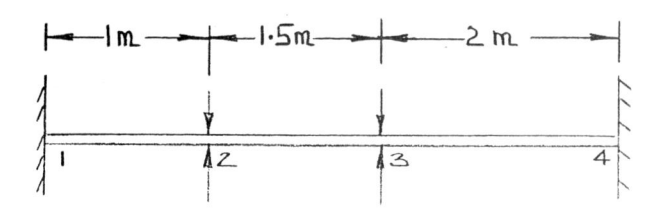

Figure 4.9 Continuous beam.

Element 1-2

From equation (2.94)

$$[k_{1-2}] = 2 \text{ E11} \times 2 \text{ E-7} \begin{matrix} \theta_2 \\ [4] \theta_2 \end{matrix}$$

$$= [0.16 \text{ E6}] \theta_2 \qquad (4.104)$$

Continuous Beams

From equation (4.98)

$$[m_{1-2}] = \frac{7860 \times 1E-3 \times 1}{420} \begin{matrix} \theta_2 \\ [4] \theta \end{matrix}$$

$$= [0.0749] \begin{matrix} \theta_2 \\ \theta_2 \end{matrix} \qquad (4.105)$$

Element 2-3

From equation (2.94)

$$[k_{2-3}] = 2E11 \times 2E-7 \begin{bmatrix} \theta_2 & \theta_3 \\ 4/1.5 & 2/1.5 \\ 2/1.5 & 4/1.5 \end{bmatrix} \begin{matrix} \theta_2 \\ \theta_3 \end{matrix}$$

$$= 1E6 \begin{bmatrix} \theta_2 & \theta_3 \\ 0.107 & 0.0533 \\ 0.0533 & 0.107 \end{bmatrix} \begin{matrix} \theta_2 \\ \theta_3 \end{matrix} \qquad (4.106)$$

From equation (4.98)

$$[m_{2-3}] = \frac{7860 \times 1E-3 \times 1.5}{420} \begin{bmatrix} 4 \times 1.5^2 & -3 \times 1.5^2 \\ -3 \times 1.5^2 & 4 \times 1.5^2 \end{bmatrix}$$

$$= \begin{bmatrix} \theta_2 & \theta_3 \\ 0.253 & -0.189 \\ -0.189 & 0.253 \end{bmatrix} \begin{matrix} \theta_2 \\ \theta_3 \end{matrix} \qquad (4.107)$$

Element 3-4

From equation (2.94)

$$[k_{3-4}] = 2E11 \times 2E-7 \begin{bmatrix} 4 \\ 2 \end{bmatrix}\begin{matrix} \theta_3 \\ \theta_3 \end{matrix}$$

$$[k_{3-4}] = [\ 0.08\ E6\ ]\ \theta_3 \qquad (4.108)$$

From equation (4.98)

$$[m_{3-4}] = \frac{7860 \times 1E-3 \times 2}{420} [\ 4 \times 2^2\ ]\ \theta_3$$

$$= [\ 0.599\ ]\ \theta_3 \qquad (4.109)$$

From equations (4.104), (4.106) and (4.108)

$$[K_{11}] = 1E6 \begin{bmatrix} 0.16 & 0.0533 \\ +0.107 & \\ \hline 0.0533 & 0.107 \\ & +0.08 \end{bmatrix} \begin{matrix} \theta_2 \\ \\ \theta_3 \end{matrix} \qquad (4.110)$$

$$= 1E6 \begin{bmatrix} 0.267 & 0.0533 \\ 0.0533 & 0.187 \end{bmatrix} \begin{matrix} \theta_2 \\ \theta_3 \end{matrix} \qquad (4.111)$$

From equations (4.105), (4.107) and (4.109)

$$[M_{11}] = \begin{bmatrix} 0.0749 \\ +0.253 & -0.189 \\ \hline -0.189 & 0.253 \\ & +0.599 \end{bmatrix} \begin{matrix} \theta_2 \\ \\ \theta_3 \end{matrix} \quad \begin{matrix} \theta_2 & \theta_3 \end{matrix} \quad (4.112)$$

$$[M_{11}] = \begin{bmatrix} 0.328 & -0.189 \\ -0.189 & 0.852 \end{bmatrix} \begin{matrix} \theta_2 \\ \theta_3 \end{matrix} \quad (4.113)$$

Now $\left| [K_{11}] - \omega^2 [M_{11}] \right| = 0$

$$\therefore \left| 1E6 \begin{bmatrix} 0.267 & 0.0533 \\ 0.0533 & 0.187 \end{bmatrix} - \omega^2 \begin{bmatrix} 0.328 & -0.189 \\ -0.189 & 0.852 \end{bmatrix} \right| = 0$$

$$\begin{vmatrix} (0.267\ E6 - 0.328\ \omega^2) & (0.0533\ E6 + 0.189\ \omega^2) \\ (0.0533\ E6 + 0.189\ \omega^2) & (0.187\ E6 - 0.852\ \omega^2) \end{vmatrix} = 0 \quad (4.114)$$

Expanding the determinant of equation (4.114)

$$(0.267\ E6 - 0.328\ \omega^2)(0.187\ E6 - 0.852\ \omega^2)$$
$$- (0.0533\ E6 + 0.189\ \omega^2)^2 = 0$$
$$4.993\ E10 + 0.279\ \omega^4 - 0.289\ E6\ \omega^2$$
$$- 0.284\ E10 - 0.036\ \omega^4 - 0.02\ E6\ \omega^2 = 0$$
$$0.243\ \omega^4 - 0.309\ E6\ \omega^2 + 4.709\ E10 = 0$$

Solving the quadratic equation (4.115)

$$\omega^2 = \frac{+0.309\ E6 \pm 0.223\ E6}{0.486}$$

$\omega_1^2 = 176955;$ $\omega_1 = 420.7$ rads/s; $\underline{n_1 = 67.0\ Hz}$

$\omega_2^2 = 1.095\ E6;$ $\omega_2 = 1046.2$ rads/s; $\underline{n_2 = 166.5\ Hz}$

4.6 Rigid-jointed Plane Frames

Prior to obtaining the mass matrix for a rigid-jointed plane frame, it will be necessary to obtain the mass matrix for the inclined beam element of Figure 4.10.

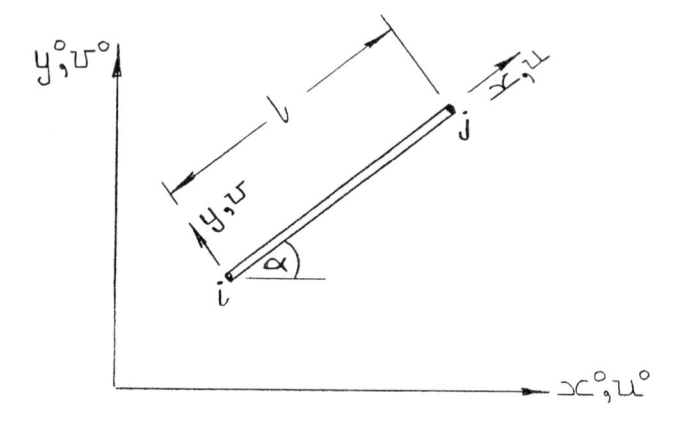

Figure 4.10 Inclined Beam Element.

In this case, there are four local degrees of freedom and four global degrees of freedom, and from equation (2.111), these are related by the following expression:

Rigid-Jointed Plane Frames

$$\begin{Bmatrix} u_i \\ v_i \\ \theta_i \\ u_j \\ v_j \\ \theta_j \end{Bmatrix} = \begin{bmatrix} \begin{array}{ccc} c & s & 0 \\ -s & c & 0 \\ 0 & 0 & 1 \end{array} & 0_3 \\ \hline 0_3 & \begin{array}{ccc} c & s & 0 \\ -s & c & 0 \\ 0 & 0 & 1 \end{array} \end{bmatrix} \begin{Bmatrix} u_i^o \\ v_i^o \\ \theta_i \\ u_j^o \\ v_j^o \\ \theta_j \end{Bmatrix} \quad (4.115)$$

where
$\quad c = \cos \alpha$
$\quad s = \sin \alpha$

$$\text{or } \{U_i\} = \begin{bmatrix} T & 0_3 \\ 0_3 & T \end{bmatrix} \{U_i^o\} \quad (4.116)$$

Now $[m_b^o]$ = the mass matrix for an inclined beam in global co-ordinates.

$$= \begin{bmatrix} T & 0_3 \\ \hline 0_3 & T \end{bmatrix}^T \begin{bmatrix} m \end{bmatrix} \begin{bmatrix} T & 0_3 \\ \hline 0_3 & T \end{bmatrix} \quad (4.117)$$

$$\therefore [m_b^o] = \frac{\rho A \ell}{420} \begin{bmatrix} 156s^2 & & & & & \\ -156cs & 156c^2 & & & & \\ 22\ell s & -22\ell c & 4\ell^2 & & & \\ 54s^2 & -54cs & 13\ell s & 156s^2 & & \\ -54cs & 54c^2 & -13\ell c & -156cs & 156c^2 & \\ -13\ell s & 13\ell c & -3\ell^2 & -22\ell s & 22\ell c & 4\ell^2 \end{bmatrix} \begin{matrix} u_i^o \\ v_i^o \\ \theta_i \\ u_j^o \\ v_j^o \\ \theta_j \end{matrix} \quad (4.118)$$

with column headings $u_i^o \quad v_i^o \quad \theta_i \quad u_j^o \quad v_j^o \quad \theta_j$

For the rigid-jointed plane frame, its mass matrix, namely [m°], is a combination of flexural and axial terms,

$$\text{ie } [m°] = [m_b°] + [m_r°] \tag{4.119}$$

where $[m_r°]$ = the x components of equation (4.29), the y components already being allowed for in the derivation of equation (4.118)

$$\text{ie } [m_r] = \frac{\rho A \ell}{6} \begin{bmatrix} 2 & 0 & 0 & 1 & 0 & 0 \\ 0 & 0 & 0 & 0 & 0 & 0 \\ 0 & 0 & 0 & 0 & 0 & 0 \\ 1 & 0 & 0 & 2 & 0 & 0 \\ 0 & 0 & 0 & 0 & 0 & 0 \\ 0 & 0 & 0 & 0 & 0 & 0 \end{bmatrix} \begin{matrix} u_i^o \\ v_i^o \\ \theta_i \\ u_j^o \\ v_j^o \\ \theta_j \end{matrix} \tag{4.120}$$

$$\text{Now } [m_r°] = \begin{bmatrix} T & 0_3 \\ \hline 0_3 & T \end{bmatrix}^T [m_r] \begin{bmatrix} T & 0_3 \\ \hline 0_3 & T \end{bmatrix}$$

$$\therefore [m_r°] = \frac{\rho A \ell}{6} \begin{bmatrix} 2c^2 & & & & & \\ 2cs & 2s^2 & & & & \\ 0 & 0 & 0 & & & \\ c^2 & cs & 0 & 2c^2 & & \\ cs & s^2 & 0 & 2cs & 2s^2 & \\ 0 & 0 & 0 & 0 & 0 & 0 \end{bmatrix} \begin{matrix} u_i^o \\ v_i^o \\ \theta_i \\ u_j^o \\ v_j^o \\ \theta_j \end{matrix} \tag{4.121}$$

Sec. 4.6]   Rigid-Jointed Plane Frames

If a concentrated mass of magnitude M_c and a mass moment of inertia of I_c is added to a node "i", the effect of this must be added to equation (4.119), as follows

$$\begin{matrix} u_i^o & v_i^o & \theta_i^o \end{matrix}$$
$$\begin{bmatrix} M_c & 0 & 0 \\ 0 & M_c & 0 \\ 0 & 0 & I_c \end{bmatrix} \begin{matrix} u_i^o \\ v_i^o \\ \theta_i^o \end{matrix} \quad \text{at node i} \tag{4.122}$$

4.6.1 Vibrations of rigid-jointed plane frames.

As each node has 3 degrees of freedom, it will be very difficult to determine by hand calculation the resonant frequencies and eigenmodes of a typical rigid-jointed plane frame. Hence, with the aid of the computer program "VIBRJPF" [1], a typical rigid-jointed plane frame will now be analysed.

Example 4.7

Determine the first six natural frequencies of vibration and eigenmodes of the rigid-jointed plane frame of Figure 4.11, given that:

$A = 0.001 \text{ m}^2$, $\quad I = 2 \times 10^{-6} \text{ m}^4$, $\quad E = 2 \times 10^{11} \text{ N/m}^2$ $\quad \rho = 7860 \text{ kg/m}^3$

It may be assumed that all the elements are of 1 m length.

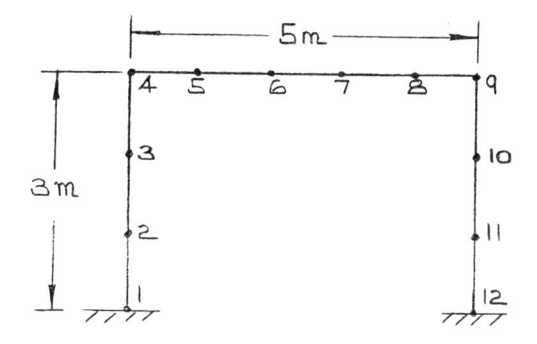

Figure 4.11 Rigid-Jointed plane frame

The results are shown in Figures 4.12 to Figure 4.17.

```
TO CONTINUE, TYPE Y
EIGENMODE= 1   FREQUENCY= 10.07721 Hz
```

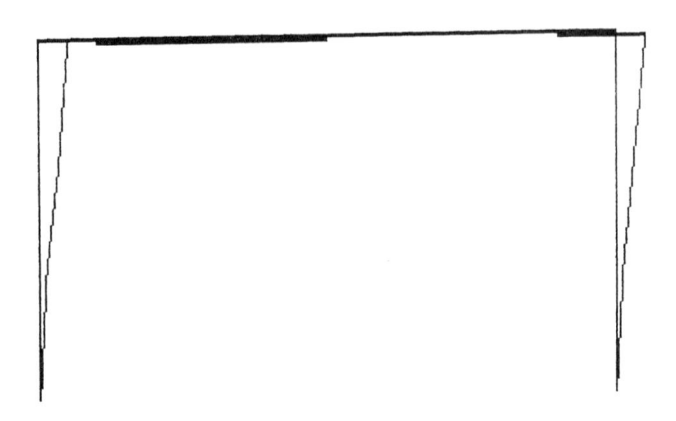

Figure 4.12 First Eigenmode.

```
TO CONTINUE, TYPE Y
EIGENMODE= 2   FREQUENCY= 22.63725 Hz
```

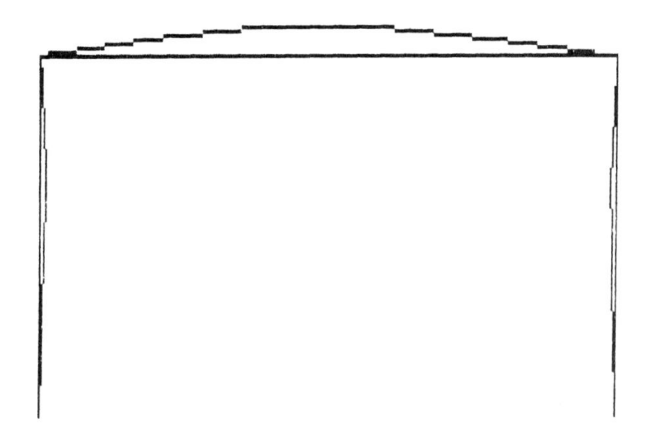

Figure 4.13 Second Eigenmode.

```
TO CONTINUE, TYPE Y
EIGENMODE= 3   FREQUENCY= 60.22066 Hz
```

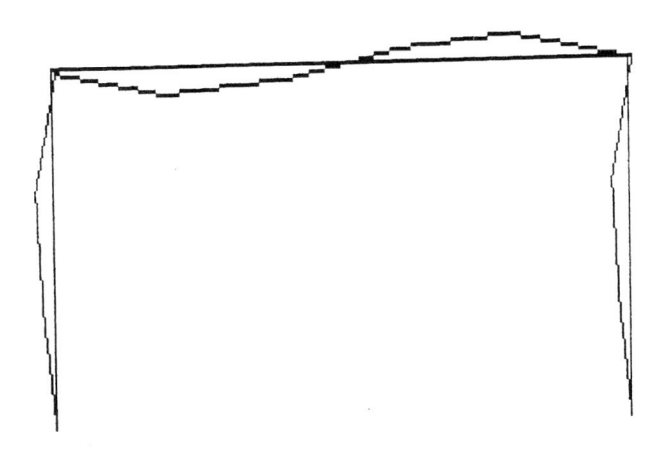

Figure 4.14 Third Eigenmode.

```
TO CONTINUE, TYPE Y
EIGENMODE= 4   FREQUENCY= 74.31848 Hz
```

Figure 4.15 Fourth Eigenmode.

```
TO CONTINUE, TYPE Y
EIGENMODE= 5   FREQUENCY= 93.2206 Hz
```

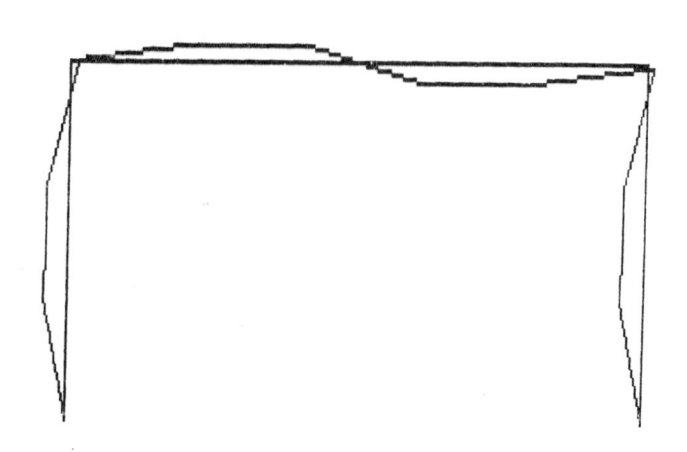

Figure 4.16 Fifth Eigenmode.

```
TO CONTINUE, TYPE Y
EIGENMODE= 6   FREQUENCY= 147.0646 Hz
```

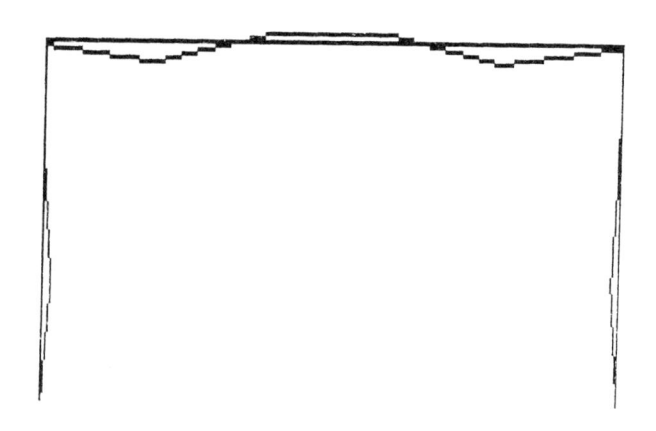

Figure 4.17 Sixth Eigenmode.

Examples for Practice 4

1. Working from first principles, determine the mass matrix for the in-plane annular element of Figure 4.18. The element may be assumed to be of uniform thickness "t" and to be described by an internal nodal circle "i" on its internal circumference and an external nodal circle "j" on its external circumference.

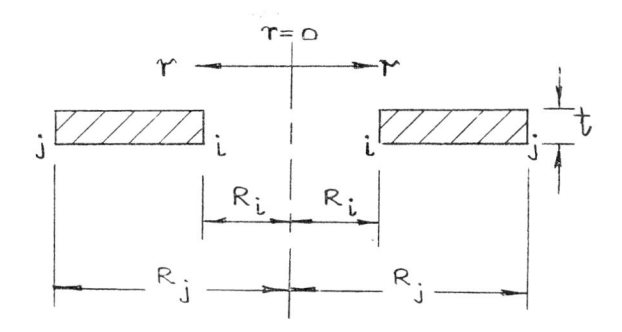

Figure 4.18 In-plane annular element.

2. Determine the natural frequency of vibration for a one element rod of length ℓ, cross-sectional area A, fixed at node 1 and free at node 2.

3. Determine the two natural frequencies of vibration and eigenmodes, for a one element rod, of length "ℓ" and cross-sectional area "A", which is free at both nodes 1 and 2.

4. Determine the two natural frequencies of vibration for a one element uniform section cantilever of length "ℓ", cross-sectional area "A" and second moment of area "I".

5. Determine the two natural frequencies of vibration and eigenmodes for the pin-jointed plane trusses of Figure 4.19(a) and 4.19(b). It may be assumed that for both trusses,

$A = 1E\text{-}3 \ m^2; \qquad E = 2 \times 10^{11} \ N/m^2; \qquad \rho = 7860 \ kg/m^3$

(a)   (b)

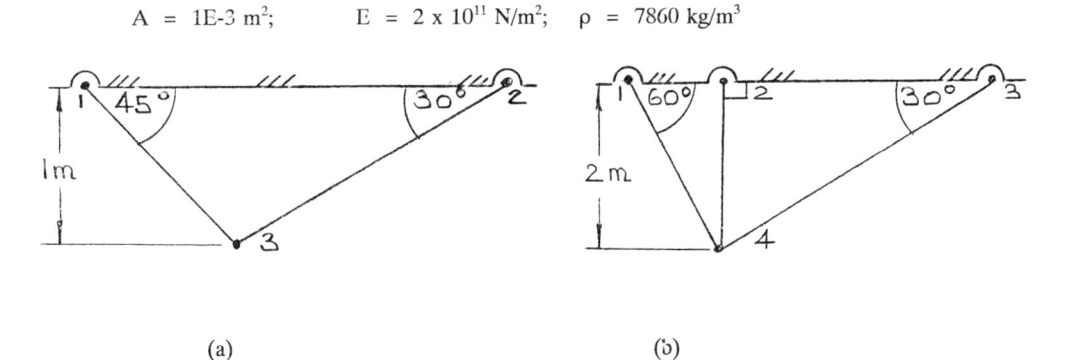

Figure 4.19 Plane pin-jointed trusses.

6. Determine the natural frequencies of vibration and eigenmodes of the pin-jointed space trusses of Figures 4.20 and 4.21, which are both securely anchored at their bases. For both cases,

$A = 0.001 \text{ m}^2;$ $E = 2 \times 10^{11} \text{ N/m}^2;$ $\rho = 7860 \text{ kg/m}^3$

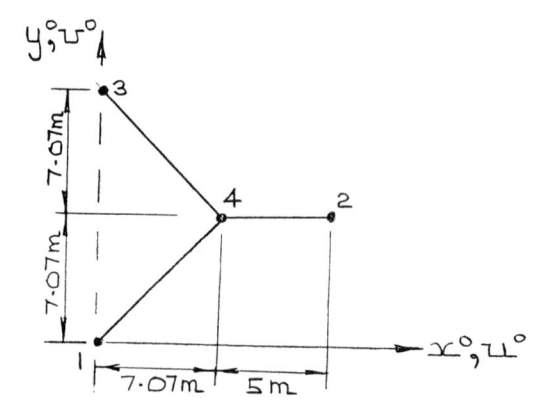

(a) Plan

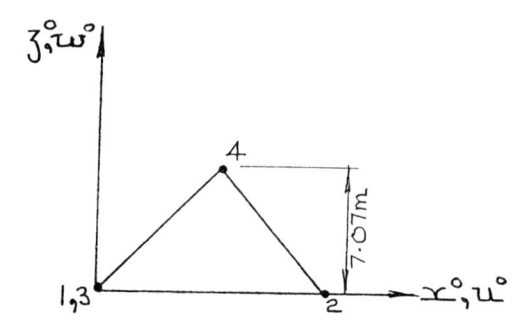

(b) Front View

Figure 4.20 Pin-jointed tripod.

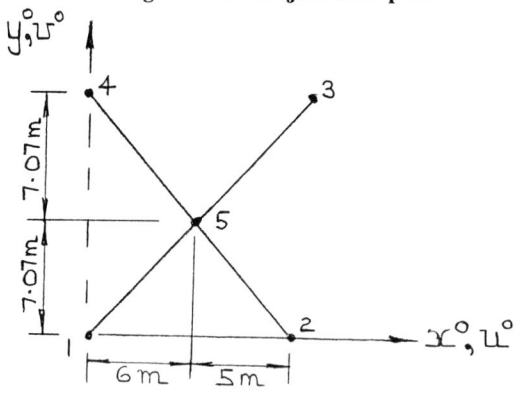

(a) Plan View

Sec. 4.6] Rigid-Jointed Plane Frames 157

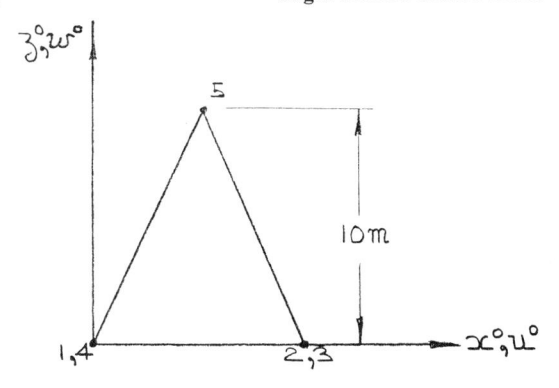

(b) Front View

Figure 4.21 Pin-jointed space truss.

7. Determine the natural frequencies of vibration and corresponding eigenmodes for the continuous beams of Figures 4.22(a) and 4.22(b), which are fixed at their extremities. It may be assumed that

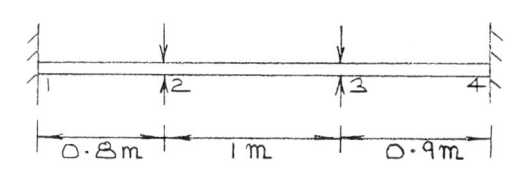

$A = 0.0006$ m^2; $I = 2 \times 10^{-7}$ m^4; $E = 1 \times 10^{11}$ N/m^2; $\rho = 2620$ kg/m^3.

(a) Beam simply supported at nodes 2 and 3.

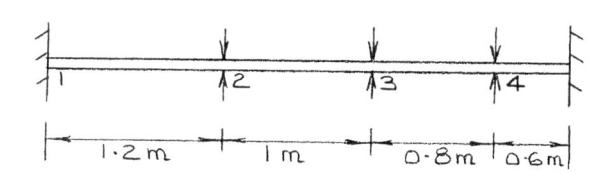

(b) Beam simply supported at nodes 2, 3 and 4.

Figure 4.22 Continuous beams.

Chapter 5

Non-Linear Structural Mechanics

5.1 Introduction

So far, we have only considered structures that deform in a linear elastic manner. That is, the load-displacement relationship of the structure behaves linearly, and also that, all the stresses in the structure remain within the elastic limit. For example, if the value of the load is doubled, the values of the displacements and stresses are also assumed to be doubled, according to linear elastic theory.

In practice, this is not always the case, as the structure could behave in a geometrically non-linear manner or in a materially non-linear manner, or some combination of the two; these non-linear analyses are now described.

5.2 Geometrical Non-Linearity

In this case, although the structure is still elastic, the structure suffers large deflections, so that its initial geometry changes. A classical case of this is that of the flat plate under lateral loading.

In this case, the lateral deflection of the flat plate causes it to become a shallow shell, so that the deformed plate resists some of its lateral load in a membrane manner, rather similar to a rubber balloon being blown up.

For geometrically non-linear problems, the structure can become stiffer if it is in tension, as it is in the case of a flat plate under lateral loading, or in the case of a rubber balloon being blown up, and it can become less stiff if it is under compression, as it is in the case of an axially loaded strut.

5.3 Material Non-Linearity

In this case, the stresses in parts of the structure exceed the yield point of the material of construction, so that the structure becomes plastic in these zones. A typical case in this category can be that of a plate with a hole in it, or other similar problem involving stress concentrations, where the material becomes plastic in the region of high stress.

In some cases, such as for frameworks, the plastic zones can behave like plastic hinges, so that the structure can fail in the manner of a mechanism; the plastic hinges acting like "rusty" hinges in a mechanism.

5.4 Combined Geometrical and Material Non-Linearity

In this case, both geometrical and material non-linearity are important, so that the stiffness matrix [K] becomes:

$$[K] = [K(u_i^o) + K_G(u_i^o)] \qquad (5.1)$$

where

$[K(u_i^o)]$ = the stiffness matrix at the ith step

$[K_G(u_i^o)]$ = the geometrical stiffness matrix at the ith step

Combined Geometrical and Material Non-Linearity

Prior to loading, or in step 0,

$$[K] = [K(0)] \quad (5.2)$$

and $\{q_i\} = [K]\{U_i\} \quad (5.3)$

After step 1,

$\{U_i\} = \{u_i^\circ\}$, and $[K_G]$ can be calculated, as it is dependent on the internal stresses in the structure. Additionally $[K(u_i^\circ)]$ can be calculated by updating the geometry of the structure, and by taking into consideration the plastic Young's modulus for any element in the structure that may have become plastic.

Thus, for problems involving material and geometrical non-linearity, it is convenient to use an incremental step-by-step analysis, as described by Table 5.1.

Step	$\{\delta q^\circ\}$	Stiffness Matrix	$\{\delta u^\circ\}$	Displacements
1	$\{\delta q_1^\circ\}$	$[K^\circ(0)] + [K_G^\circ(0)]$ $\{\delta u_1^\circ\}$		$\{u_1^\circ\} = \{\delta u_1^\circ\}$
2	$\{\delta q_2^\circ\}$	$[K^\circ(u_1^\circ)] + [K_G^\circ(u^\circ)]$	$\{\delta u_2^\circ\}$	$\{u_2^\circ\} = \{u_1^\circ\} + \{\delta u_2^\circ\}$
3	$\{\delta q_3^\circ\}$	$[K^\circ(u_2^\circ)] + [K_G^\circ(u_2^\circ)]$	$\{\delta u_3^\circ\}$	$\{u_3^\circ\} = \{u_2^\circ\} + \{\delta u_3^\circ\}$
.	.	.	.	.
.	.	.	.	.
.	.	.	.	.
n	$\{\delta q_n^\circ\}$	$[K^\circ(u_{n-1}^\circ)] + [K_G^\circ(u_{n-1}^\circ)]$	$\{\delta u_n^\circ\}$	$\{u_n^\circ\} = \{u_{n-1}^\circ\} + \{\delta u_n^\circ\}$
Σ	$\{q_n^\circ\}$		$\{u_n^\circ\}$	

Table 5.1 Incremental Step by Step Method.

In Table 5.1, the process is to calculate the deflections and stresses from equation (5.3), due to a small load $\{\delta q_i^\circ\}$ and to up date the geometry of the structure, so that $[K(u_1^\circ)]$ can be calculated, making allowances for any loss in value for Young's modulus, in any element due to plastic deformation.

Additionally, as the structure now has internal stresses, $[K_G]$ can be calculated, remembering that a "negative" $[K_G]$ will decrease the stiffness of the structure and a "positive" $[K_G]$ will increase the stiffness of the structure, as shown by equation (5.1).

The process now is to add an additional load for the second step, namely, $\{\delta q_2^\circ\}$, and to calculate the resulting deflections $\{\delta u_2^\circ\}$ and stresses, both of which must be added to the corresponding deflections and stresses obtained after the first step. The geometry of the structure must now be updated and any effects due to plasticity, must be taken into consideration in calculating $[K(u_2^\circ)]$. As the values of the internal stresses have now changed, $[K_G]$ can be recalculated. The process is then repeated in the suitable increments, until eventually, the structure collapses. However, it should be emphasised that in some cases, convergence can be difficult.

5.5 Problems Involving Geometric Non-Linearity

For purely geometrically non-linear problems, the method described in Section 5.4 can be used, except that the effects of material non-linearity should not be included.

In this section, however, we will restrict ourselves to elastic instability problems, similar to those covered by the Euler theory of buckling.

In Table 5.1, if there are no internal stresses, so that $[K^o{}_G(0)]$ is null, then the load-displacement relationship is given by:

$$\{U_i\} = [K^o]^{-1} \{q_i\}, \tag{5.4}$$

where

$\{q_i\}$ = the vector of forces to cause elastic buckling; this is unknown.

If, however, we assume that

$$\{q_i\} = \lambda \{q^*_i\}, \tag{5.5}$$

where λ = a constant

and $\{q^*_i\}$ = a vector containing the relative magnitudes of the externally applied forces to cause elastic instability.

Hence, equation (5.4) becomes

$$\{U_i\} = [K^o]^{-1} \lambda \{q^*_i\} \tag{5.6}$$

Letting $\lambda = 1$ in equation (5.6)

$$\{U^*_i\} = [K^o]^{-1} \{q^*_i\} \tag{5.7}$$

Hence, $[K^*_G]$ can be determined,

where $[K_G] = \lambda [K^*_G]$ \hfill (5.8)

Now as elastic instability problems are linearly elastic,

$[K^o]$ will not change, so that

$$\{q_i\} = ([K^o] + \lambda [K^*_G]) \{U_i\} \tag{5.9}$$

When elastic buckling is about to occur,

$$\{q_i\} = \{q_{cr}\} \tag{5.10}$$

where

$\{q_{cr}\}$ = the vector of loads required to cause elastic buckling.

Additionally, $\{U_i\} \rightarrow \{\infty\}$,

hence,

$$| [K^o] + \lambda [K^*_G] | = 0 \tag{5.11}$$

For constrained structures,

$$| [K_{11}] + \lambda [K_{G11}] | = 0 \tag{5.12}$$

where

$$[K_{G11}] = \Sigma[k_G] \qquad (5.13)$$

= that part of the system geometrical stiffness matrix that corresponds to the free displacements

$[k_G]$ = elemental geometrical stiffness matrix

$$\{q_{cr}\} = \lambda_{cr} \{q^*_{cr}\} \qquad (5.14)$$

λ_{cr} = the minimum value of λ, in magnitude.

5.5.1 Geometrical Stiffness Matrix for a Rod Element

To determine the geometrical stiffness matrix for a rod element, the additional effects of large deflections have to be taken into account. Although these effects are in the axial direction of the rod element, they are caused by lateral deformations in the element, just prior to buckling, as shown in Figure 5.1.

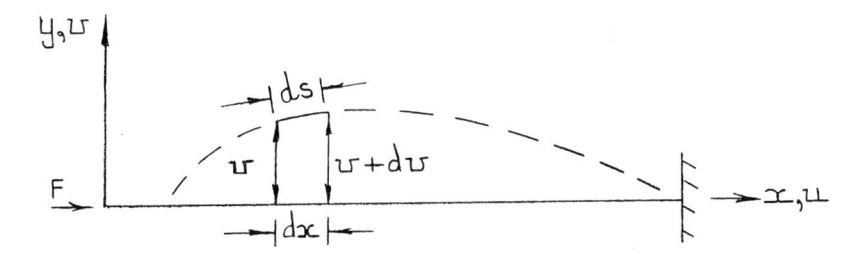

Figure 5.1. Deformed shape of the rod element just prior to buckling.

From Figure 5.1 and by Pythagoras' theorem,

$$(\delta s)^2 = (\delta v)^2 + (\delta x)^2 \qquad (5.15)$$

$$\text{or } \delta s = \left[1 + \left(\frac{\delta v}{\delta x}\right)^2\right]^{\frac{1}{2}} \delta x \qquad (5.16)$$

Expanding equation (5.16) binomially, we get,

$$\frac{\delta s}{\delta x} = 1 + \frac{1}{2}\left(\frac{\delta v}{\delta x}\right)^2 - \frac{1}{8}\left(\frac{\delta v}{\delta x}\right)^4 + \ldots \qquad (5.17)$$

Neglecting higher order terms in equation (5.17), we get,

$$\frac{ds}{dx} = 1 + \frac{1}{2}\left(\frac{dv}{dx}\right)^2 \qquad (5.18)$$

Hence the additional strain in the axial direction, due to $\left(\frac{dv}{dx}\right)$ is given by

$$\delta\varepsilon_x = \frac{1}{2}\left(\frac{dv}{dx}\right)^2, \qquad (5.19)$$

So that the total strain in the x direction is given by

$$\varepsilon_x = \frac{du}{dx} + \frac{1}{2}\left(\frac{dv}{dx}\right)^2 \qquad (5.20)$$

Now for the rod element of Figure 5.2, it is reasonable to assume that the element will deform as shown.

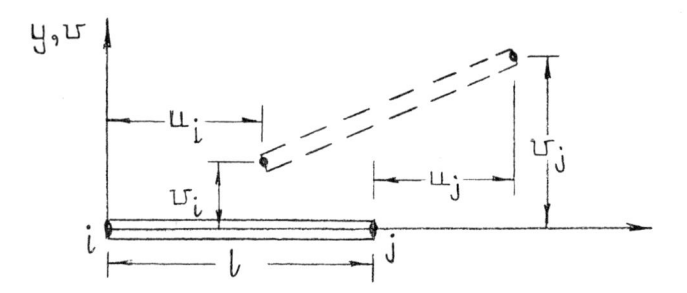

Figure 5.2. Large deflection form of rod element.

The following displacement functions, (see Chapter 3) for u and v are convenient:

$$u = u_i(1 - x/\ell) + u_j x/\ell \qquad (5.21)$$

$$v = v_i(1 - x/\ell) + v_j x/\ell \qquad (5.22)$$

Now the strain energy (U_e) stored in a rod is given by

$$U_e = \frac{AE}{2}\int_0^\ell \varepsilon_x^2 \, dx \qquad (5.23)$$

Problems Involving Geometric Non-Linearity

Hence from equation (5.23)

$$U_e = \frac{AE}{2} \int_0^\ell \left[\left(\frac{du}{dx}\right)^2 + \frac{du}{dx}\left(\frac{dv}{dx}\right)^2 + \frac{1}{4}\left(\frac{dv}{dx}\right)^4 \right] dx \quad (5.24)$$

Neglecting higher order terms,

$$U_e = \frac{AE}{2} \int_0^\ell \left[\left(\frac{du}{dx}\right)^2 + \frac{du}{dx}\left(\frac{dv}{dx}\right)^2 \right] dx \quad (5.25)$$

Substituting equations (5.21) and (5.22) into equation 5.25).

$$U_e = \frac{AE}{2} \int_0^\ell \left[\left(\frac{-u_i + u_j}{\ell}\right)^2 + \left(\frac{-u_i + u_j}{\ell}\right)\left(\frac{-v_i + v_j}{\ell}\right)^2 \right] dx$$

$$= \frac{AE}{2\ell}\left(u_i^2 - 2u_i u_j + u_j^2\right) + \frac{AE}{2\ell^2}\left(u_j - u_i\right)\left(v_i^2 - 2v_i v_j + v_j^2\right)$$

$$= \frac{AE}{2\ell}\left(u_i^2 - 2u_i u_j + u_j^2\right) + \frac{F}{2\ell}\left(v_i^2 - 2v_i v_j + v_j^2\right) \quad (5.26)$$

where

$$F = \frac{AE}{\ell}\left(u_j - u_i\right) \quad (5.27)$$

Applying the method of minimum potential to equation (5.26), the elemental stiffness matrix can be obtained from equations (5.28), as follows,

$$\frac{\partial U_e}{\partial u_i} = \frac{AE}{\ell}\left(u_i - u_j\right)$$

$$\frac{\partial U_e}{\partial u_j} = \frac{AE}{\ell}\left(-u_i + u_j\right)$$

$$\frac{\partial U_e}{\partial v_i} = \frac{F}{\ell}\left(v_i - v_j\right) \quad (5.28)$$

$$\frac{\partial U_e}{\partial v_j} = \frac{F}{\ell}\left(-v_i + v_j\right)$$

Rewriting equations (5.28) in matrix form, the following relationship is obtained:

$$[k]\,\{U_i\} = \left(\frac{AE}{\ell} \begin{bmatrix} 1 & 0 & -1 & 0 \\ 0 & 0 & 0 & 0 \\ -1 & 0 & 1 & 0 \\ 0 & 0 & 0 & 0 \end{bmatrix} + \frac{F}{\ell} \begin{bmatrix} 0 & 0 & 0 & 0 \\ 0 & 1 & 0 & -1 \\ 0 & 0 & 0 & 0 \\ 0 & -1 & 0 & 1 \end{bmatrix} \right) \begin{Bmatrix} u_i \\ v_i \\ u_j \\ v_j \end{Bmatrix} \tag{5.29}$$

where

$$[k] = \frac{AE}{\ell} \begin{array}{c} \begin{matrix} u_i & v_i & u_j & v_j \end{matrix} \\ \begin{bmatrix} 1 & 0 & -1 & 0 \\ 0 & 0 & 0 & 0 \\ -1 & 0 & 1 & 0 \\ 0 & 0 & 0 & 0 \end{bmatrix} \end{array} \begin{matrix} u_i \\ v_i \\ u_j \\ v_j \end{matrix} \tag{5.30}$$

= the elemental stiffness matrix for a rod in local co-ordinates.

$$\text{and } [k_G] = \frac{F}{\ell} \begin{array}{c} \begin{matrix} u_i & v_i & u_j & v_j \end{matrix} \\ \begin{bmatrix} 0 & 0 & 0 & 0 \\ 0 & 1 & 0 & -1 \\ 0 & 0 & 0 & 0 \\ 0 & -1 & 0 & 1 \end{bmatrix} \end{array} \begin{matrix} u_i \\ v_i \\ u_j \\ v_j \end{matrix} \tag{5.31}$$

= the geometrical stiffness matrix for a rod in local co-ordinates.

For the buckling of pin-jointed trusses, the geometrical stiffness matrix in global co-ordinates $[k_G^o]$ is required, this can be obtained from equation (5.32), as follows:

$$[k_G^o] = \begin{bmatrix} T & 0_2 \\ \hline 0_2 & T \end{bmatrix}^T [k_G] \begin{bmatrix} T & 0_2 \\ \hline 0_2 & T \end{bmatrix} \quad (5.32)$$

where

$$[T] = \begin{bmatrix} c & s \\ -s & c \end{bmatrix} \quad (5.33)$$

$c = \cos \alpha$

$s = \sin \alpha$

$\alpha = $ angle of inclination of a rod element from the horizontal, (positive is counter-clockwise); see Figure 5.3.

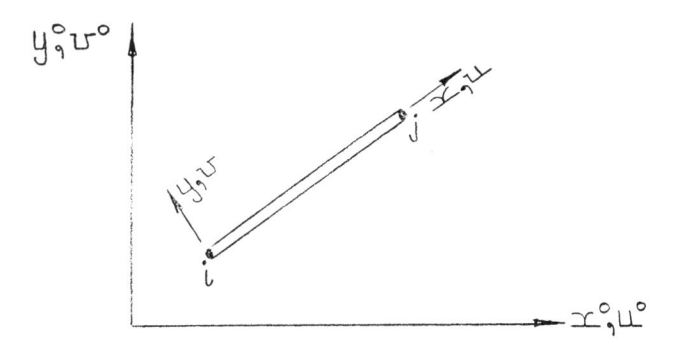

Figure 5.3. Rod element in global co-ordinates.

Substituting equations (5.31) and (5.33) into equation (5.32), the geometrical stiffness matrix for a rod element in global co-ordinates, $[k_G^o]$, is given by:

$$[k_G^o] = \frac{F}{\ell} \begin{bmatrix} s^2 & -cs & -s^2 & cs \\ -cs & c^2 & cs & -c^2 \\ -s^2 & cs & s^2 & -cs \\ cs & -c^2 & -cs & c^2 \end{bmatrix} \quad (5.34)$$

5.5.2 Geometrical Stiffness Matrix for a Beam Element

This element is useful for the non-linear analysis of axially loaded struts and ties and for rigid-jointed plane frames. The axially loaded struts and ties can be accompanied by a complex combination of lateral loads and couples.

From Chapter 3, the displacement functions for the beam in flexure, shown in Figure 5.4, are as follows:

$$u = u_i(1 - \xi) + u_j\xi \qquad (5.35)$$

and

$$v = v_i(1 - 3\xi^2 + 2\xi^3) + \theta_i \ell(-\xi + 2\xi^2 - \xi^3) + v_j(3\xi^2 - 2\xi^3)$$

$$+ \theta_j \ell(\xi^2 - \xi^3), \qquad (5.36)$$

where

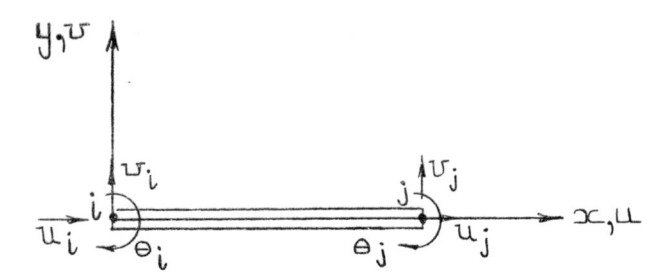

$$\xi = x/\ell \qquad (5.37)$$

Figure 5.4 Beam-column.

Now from reference [4] and section 5.5.1, it can be readily shown that the strain in the fibre of a beam under combined axial and bending load is given by:

$$\varepsilon_x = \frac{du}{dx} + \frac{1}{2}\left(\frac{dv}{dx}\right)^2 - \frac{d^2v}{dx^2} \cdot y, \qquad (5.38)$$

where

$$\frac{1}{2}\left(\frac{dv}{dx}\right)^2 = \text{additional axial strain due to large deflections,}$$

and

$$\frac{-d^2v}{dx^2} \cdot y = \text{the axial strain due to bending, in a fibre, at a distance}$$

"y" from the neutral axis.

Now the total strain energy (U_e) in the beam-column is given by:

$$U_e = \frac{E}{2} \int_{vol} \varepsilon_x^2 \, d(vol) \tag{5.39}$$

Substituting the appropriate derivatives of equations (5.35) and (5.36) into equation (5.39), and by applying the method of minimum potential, but neglecting the term $(dv/dx)^4$, it can be show that the stiffness matrix for a beam-column can be obtained from the following expression:

$$\left([k] + [k_G]\right)\{U_i\} = \frac{EI}{\ell^3}\begin{bmatrix} A\ell^2/I & & & & & \\ 0 & 12 & & \text{symmetrical} & & \\ 0 & -6\ell & 4\ell^2 & & & \\ -A\ell^2/I & 0 & 0 & A\ell^2/I & & \\ 0 & -12 & 6\ell & 0 & 12 & \\ 0 & -6\ell & 2\ell^2 & 0 & 6\ell & 4\ell^2 \end{bmatrix}$$

$$+ \frac{F}{\ell}\begin{bmatrix} 0 & & & & & \\ 0 & 6/5 & & \text{symmetrical} & & \\ 0 & -\ell/10 & 2\ell^2/15 & & & \\ 0 & 0 & 0 & 0 & & \\ 0 & -6/5 & \ell/10 & 0 & 6/5 & \\ 0 & -\ell/10 & -\ell^2/30 & 0 & \ell/10 & 2\ell^2/15 \end{bmatrix}\begin{Bmatrix} u_i \\ v_i \\ \theta_i \\ u_j \\ v_j \\ \theta_j \end{Bmatrix} \tag{5.40}$$

where

$$F = AE(u_j - u_i)/\ell$$

$[k_G]$ = geometrical stiffness matrix for a beam in local co-ordinates.

For the non-linear analysis of rigid-jointed plane frames, the geometrical stiffness matrix in global co-ordinates, namely, $[k^o{}_G]$, can be obtained from equation (5.41).

$$[k^o{}_G] = \begin{bmatrix} T & 0_3 \\ \hline 0_3 & T \end{bmatrix}^T [k_G] \begin{bmatrix} T & 0_3 \\ \hline 0_3 & T \end{bmatrix} \qquad (5.41)$$

where

$$[T] \begin{bmatrix} c & s & 0 \\ -s & c & 0 \\ 0 & 0 & 1 \end{bmatrix} \qquad (5.42)$$

where c and s have been defined in Section 5.5.1.

Hence, for a beam column in global co-ordinates, the geometrical stiffness matrix is given by:

$$[k^o_G] = \frac{F}{\ell} \begin{bmatrix} 6s^2/5 & & & & & \\ -6cs/5 & 6c^2/5 & & \text{symmetrical} & & \\ \ell s/10 & -\ell c/10 & 2\ell^2/15 & & & \\ -6s^2/5 & 6cs/5 & -\ell s/10 & 6s^2/5 & & \\ 6cs/5 & 6c^2/5 & \ell c/10 & -6cs/5 & 6c^2/5 & \\ \ell s/10 & -\ell c/10 & -\ell^2/30 & -\ell s/10 & \ell c/10 & 2\ell^2/15 \end{bmatrix} \begin{matrix} u_i^o \\ v_i^o \\ \theta_i \\ u_j^o \\ v_j^o \\ \theta_j \end{matrix} \qquad (5.43)$$

with column headings $u_i^o \quad v_i^o \quad \theta_i \quad u_j^o \quad v_j^o \quad \theta_j$

5.6 Problems Involving Elastic Buckling

In this section, two worked examples will be presented, which involve structural buckling.

Example 5.1

Determine the load "W", that will cause elastic buckling of the pin-jointed truss of Figure 5.5. It may be assumed that AE = 6

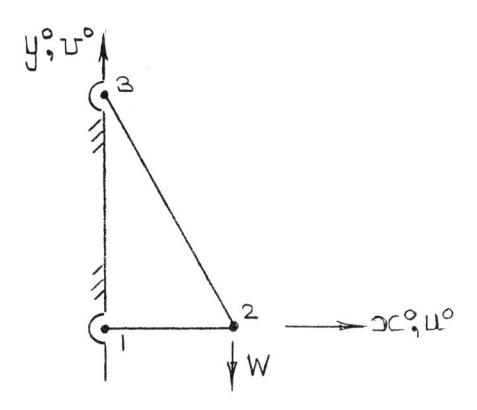

Figure 5.5 Plane Pin-jointed Truss.

Element 1-2

$\ell_{1-2} = \ell;$ $\alpha = 0;$ $c = 1;$ $s = 0$

From equation (2.43), the stiffness matrix for this element is given by:

$$[k^o_{1-2}] = \frac{6}{\ell} \begin{bmatrix} & & & \\ & & & \\ & & 1 & 0 \\ & & 0 & 0 \end{bmatrix} \begin{matrix} u_1^o \\ v_1^o \\ u_2^o \\ v_2^o \end{matrix} \quad\quad (5.44)$$

with column headers $u_1^o\; v_1^o\; u_2^o\; v_2^o$

From elementary statics [4],

F_{2-3} = force in element 2-3, due to W,

= W/sin 60°

= 1.155 W (tension)   (5.45)

and $\bar{F}_{1-2}$ = force in element 1-2, due to W,

= $-F_{2-3}$ cos 60

= -0.5775 W (compression)   (5.46)

Hence, from equation (5.34), the geometrical stiffness matrix for this element is given by:

$$[k^o_{G1-2}] = \frac{-0.5775W}{\ell} \begin{bmatrix} & & & \\ & & & \\ & & 0 & 0 \\ & & 0 & 1 \end{bmatrix} \begin{matrix} u_1^o \\ v_1^o \\ u_2^o \\ v_2^o \end{matrix} \quad\quad u_1^o\ v_1^o\ u_2^o\ v_2^o$$

(5.47)

Element 2-3

$\ell_{2-3} = \ell/\cos 60 = 2\ell;$ $\alpha = 120°;$ c = -0.5;   s = 0.866

From equation (2.43), the elemental stiffness matrix is given by:

$$[k^o_{2-3}] = \frac{6}{2\ell} \begin{bmatrix} 0.25 & -0.433 & & \\ -0.433 & 0.75 & & \\ & & & \\ & & & \end{bmatrix} \begin{matrix} u_2^o \\ v_2^o \\ u_3^o \\ v_3^o \end{matrix} \quad\quad u_2^o\ \ v_2^o\ u_3^o\ v_3^o$$

(5.48)

From equation (5.34), the geometrical stiffness matrix for this element is given by:

$$[k^o_{G2-3}] = \frac{1.155W}{2\ell} \begin{bmatrix} & u_2^o & v_2^o & u_3^o & v_3^o \\ 0.75 & 0.433 & & \\ 0.433 & 0.25 & & \\ \hline & & & \\ \hline & & & \end{bmatrix} \begin{matrix} u_2^o \\ v_2^o \\ u_3^o \\ v_3^o \end{matrix} \quad (5.49)$$

From equations (5.44) and (5.48), the system stiffness matrix corresponding to the free displacements, which in this case are u^o_2 and v^o_2, is given by:

$$[K_{11}] = \frac{1}{\ell} \begin{bmatrix} u_2^o & v_2^o \\ 6 + 0.75 & 0 - 1.299 \\ 0 - 1.299 & 0 + 2.25 \end{bmatrix} \begin{matrix} u_2^o \\ v_2^o \end{matrix} \quad (5.50)$$

$$= \frac{1}{\ell} \begin{bmatrix} u_2^o & v_2^o \\ 6.75 & -1.299 \\ -1.299 & 2.25 \end{bmatrix} \begin{matrix} u_2^o \\ v_2^o \end{matrix} \quad (5.51)$$

From equations (5.47) and (5.49), the system geometrical stiffness matrix corresponding to the free displacements, which in this case are u^o_2 and v^o_2, is given by:

$$[K_{G11}] = \frac{W}{\ell} \begin{bmatrix} u_2^o & v_2^o \\ 0 + 0.433 & 0 + 0.25 \\ 0.25 & -0.5775 + 0.144 \end{bmatrix} \begin{matrix} u_2^o \\ v_2^o \end{matrix} \quad (5.52)$$

$$= \frac{W}{\ell} \begin{bmatrix} 0.433 & 0.25 \\ 0.25 & -0.434 \end{bmatrix} \begin{matrix} u_2^o \\ \\ v_2^o \end{matrix} \qquad (5.53)$$

with u_2^o, v_2^o labels above.

From equation (5.12)

$$\left| [K_{11}] + [K_{G11}] \right| = 0$$

or

$$\left| \frac{1}{\ell} \begin{bmatrix} 6.75 & -1.299 \\ -1.299 & 2.25 \end{bmatrix} + \frac{W}{\ell} \begin{bmatrix} 0.433 & 0.25 \\ 0.25 & -0.434 \end{bmatrix} \right| = 0, \qquad (5.54)$$

or

$$\left| \begin{bmatrix} (6.75 + 0.433\ W) & (-1.299 + 0.25\ W) \\ (-1.299 + 0.25\ W) & (2.25 - 0.434\ W) \end{bmatrix} \right| = 0 \qquad (5.55)$$

Expanding the determinant of equation (5.55), we get

$$(6.75 + 0.433\ W)(2.25 - 0.434\ W) - (-1.299 + 0.25\ W)^2 = 0 \qquad (5.56)$$

or $15.19 - 2.93\ W + 0.974\ W - 0.188\ W^2 - 1.687 + 0.65\ W - 0.0625\ W^2 = 0$

or $-0.25\ W^2 - 1.306\ W + 13.503 = 0$

Solving the above quadratic equation, we get

$$W = \frac{1.306 \pm \sqrt{[1.706 + 13.503]}}{-0.5} = \frac{1.306 \pm 3.9}{-0.5}$$

$\therefore\ W = \underline{5.188\ MN} = \underline{\text{elastic instability load}}$

Example 5.2

Determine the axial buckling load for a strut, of length ℓ, which is pinned at both ends. The strut is shown in Figure 5.6.

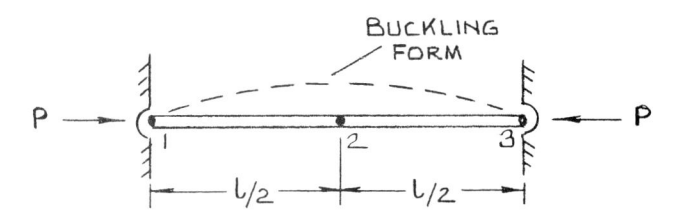

Figure 5.6. Axially loaded strut, pinned at both ends.

As the strut buckles symmetrically about its centre, we need only consider one half of the strut; in this case we can assume that $u^o_1 = v^o_1 = \theta_2 = u^o_2 = 0$

From equation (2.94) or equation (5.40),

$$[K_{11}] = \frac{EI}{(\ell/2)^3} \begin{bmatrix} 4 \times \left(\frac{\ell}{2}\right)^2 & 6 \times \left(\frac{\ell}{2}\right) \\ 6 \times \left(\frac{\ell}{2}\right) & 12 \end{bmatrix} \begin{matrix} \theta_1 \\ v^o_2 \end{matrix} \quad (5.57)$$

$$= \frac{8EI}{\ell^3} \begin{bmatrix} \ell^2 & 3\ell \\ 3\ell & 12 \end{bmatrix}$$

$$= 8EI \begin{bmatrix} 1/\ell & 3/\ell^2 \\ 3/\ell^2 & 12/\ell^3 \end{bmatrix} \begin{matrix} \theta_1 \\ v^o_2 \end{matrix} \quad (5.58)$$

From equation (5.40)

$$[K_{G11}] = \frac{P}{(\ell/2)} \begin{bmatrix} 2(\ell/2)^2/15 & (\ell/2)/10 \\ (\ell/2)/10 & 6/5 \end{bmatrix} \begin{matrix} \theta_1 \\ v_2^o \end{matrix} \quad (5.59)$$

$$= \frac{2P}{\ell} \begin{bmatrix} 2\ell^2/60 & \ell/20 \\ \ell/20 & 6/5 \end{bmatrix} \begin{matrix} \theta_1 \\ v_2^o \end{matrix} \quad (5.60)$$

$$= P \begin{bmatrix} \ell/15 & 1/10 \\ 1/10 & 12/(5\ell) \end{bmatrix} \begin{matrix} \theta_1 \\ v_2^o \end{matrix} \quad (5.61)$$

Now $\left| [K_{11}] + [K_{G11}] \right| = 0$

Hence, from equation (5.58) and (5.61)

$$\left| 8EI \begin{bmatrix} 1/\ell & 3/\ell^2 \\ 3/\ell^2 & 12/\ell^3 \end{bmatrix} + P \begin{bmatrix} \ell/15 & 1/10 \\ 1/10 & 12/(5\ell) \end{bmatrix} \right| = 0$$

$$\left| \begin{matrix} (8EI/\ell + P\ell/15) & (24EI/\ell^2 + P/10) \\ (24EI/\ell^2 + P/10) & (96EI/\ell^3 + 12P/5\ell) \end{matrix} \right| = 0 \quad (5.62)$$

Expanding equation (5.62)

$$(8EI/\ell + P\ell/15)(96 EI/\ell^3 + 12 P/5\ell)$$

$$-(24 EI/\ell^2 + P/10)^2 = 0$$

or $768 E^2 I^2/\ell^4 + 19.2 PEI/\ell^2 + 6.4 PEI/\ell^2$

$+0.16 P^2 - 576 E^2 I^2/\ell^4 - P^2/100 - 4.8 PEI/\ell^2 = 0$

or $0.15 P^2 + 20.8 PEI/\ell^2 + 192 E^2 I^2/\ell^4 = 0$

$$\therefore P = \left\{ \frac{-20.8 \pm \sqrt{432.64 - 115.2}}{0.3} \right\} \frac{EI}{\ell^2}$$

$P = -9.94 EI/\ell^2 =$ elastic instability load

NB The exact value for the elastic buckling of this strut is given by:

$$P = \frac{-\pi^2 EI}{\ell^2} = -9.87 EI/\ell^2$$

ie the error by the finite element method is +0.71%.

The error would have been a lot less if more elements were used to model the strut.

5.7 Non-linear Vibrations

If the tension in a violin string or rubber band is increased, the natural frequencies of vibration increase with increased tension [3]. Similarly, if the axially applied compressive load in a strut is increased, the natural frequencies of vibration decrease. The reason for this is that the axial tension in a rubber band causes its bending stiffness to increase, and the compressive axial load in a strut causes its bending stiffness to decrease; this behaviour will be examined with the aid of the following example.

Example 5.3

Determine the natural frequencies of a yacht mast of length 3m, under the following compressive axial forces:

(a) A compressive axial force of 1000 N.

(b) A compressive axial force of 3000 N.

(c) Zero axial force.

The following may be assumed:

$E = 1 \times 10^{10}$ N/m²

$I = 1.179 \times 10^{-6}$ m²

$A = 3.85 \times 10^{-3}$

$\rho = 600$ kg/m³

The strut is shown in Figure 5.7

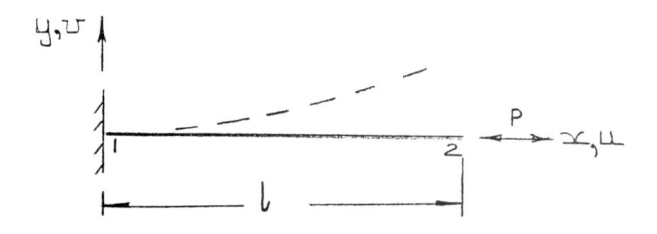

Figure 5.7 Yacht Mast.

From equation (2.94),

$$[K_{11}] = \frac{EI}{\ell^3} \begin{bmatrix} 12 & 6\ell \\ 6\ell & 4\ell^2 \end{bmatrix} \begin{matrix} v_2 \\ \theta_2 \end{matrix}$$

From equation (4.102)

(a) For a compressive axial force of 1000 N

$$[K_{11}] = \begin{matrix} v_2 & \theta_2 \end{matrix} \\ \begin{bmatrix} 5240 & 7860 \\ 7860 & 15720 \end{bmatrix} \begin{matrix} v_2 \\ \theta_2 \end{matrix} \quad (5.63)$$

$$[M_{11}] = \frac{\rho A \ell}{420} \begin{bmatrix} 156 & 22\ell \\ 22\ell & 4\ell^2 \end{bmatrix} \begin{matrix} v_2 \\ \theta_2 \end{matrix}$$

$$= \begin{bmatrix} 2.574 & 1.089 \\ 1.089 & 0.594 \end{bmatrix} \begin{matrix} v_2 \\ \theta_2 \end{matrix} \quad (5.64)$$

From equation (5.40),

$$[K_{G11}] = \frac{-1000}{3} \begin{bmatrix} 6/5 & \ell/10 \\ \ell/10 & 2\ell^2/15 \end{bmatrix} \begin{matrix} v_2 \\ \theta_2 \end{matrix}$$

$$[K_{G11}] = \begin{bmatrix} -400 & -100 \\ -100 & -400 \end{bmatrix} \begin{matrix} v_2 \\ \theta_2 \end{matrix}$$

For this case, equation (4.44) becomes

$$\left| \left([K_{11}] + [K_{G11}]\right) - \omega^2 [M_{11}] \right| = 0 \quad (5.65)$$

$$\left| \begin{bmatrix} 4840 & 7760 \\ 7760 & 15320 \end{bmatrix} - \omega^2 \begin{bmatrix} 2.574 & 1.089 \\ 1.089 & 0.594 \end{bmatrix} \right| = 0 \qquad (5.66)$$

Expanding equation (5.66), we get

$$(4840 - 2.574\,\omega^2)(15320 - 0.594\,\omega^2)$$

$$-(7760 - 1.089\,\omega^2)^2 = 0$$

or $7.415\,E7 - 2875\,\omega^2 - 39434\,\omega^2 + 1.529\,\omega^4$

$-6.022E7 + 16901\,\omega^2 - 1.186\,\omega^4 = 0$

or $0.343\,\omega^4 - 25408\,\omega^2 + 1.393\,E7 = 0 \qquad (5.67)$

Solving the above quadratic equation, we get,

$$\omega^2 = \frac{25408 \pm 25029}{0.686}$$

giving

$\omega_1^2 = 552.4;\quad \omega_1 = 23.5\text{ rads/s};\quad n_1 = 3.74\text{ Hz}$

$\omega_2^2 = 73523;\quad \omega_2 = 271.2\text{ rads/s};\quad n_2 = 43.16\text{ Hz}$

(b)   For a compressive axial force of 3000 N

From equation (5.40),

$$[K_{G11}] = \frac{-3000}{3} \begin{bmatrix} 6/5 & \ell/10 \\ \ell/10 & 2\ell^2/15 \end{bmatrix} \qquad (5.68)$$

$$= \begin{bmatrix} -1200 & -300 \\ -300 & -1200 \end{bmatrix} \qquad (5.69)$$

Hence, equation (4.44) becomes

Sec. 5.7]					Non-Linear Vibrations					179

$$\left| \begin{bmatrix} 4040 & 7560 \\ 7560 & 14520 \end{bmatrix} - \omega^2 \begin{bmatrix} 2.574 & 1.089 \\ 1.089 & 0.594 \end{bmatrix} \right| = 0 \qquad (5.70)$$

Expanding equation (5.70),

$$(4040 - 2.574 \, \omega^2)(14520 - 0.594 \, \omega^2)$$

$$-(7560 - 1.089 \, \omega^2)^2 = 0$$

or $5.866 \, E7 - 2400 \, \omega^2 - 37374 \, \omega^2 + 1.529 \, \omega^4$

$$-5.715 \, E7 + 16466 \, \omega^2 - 1.186 \, \omega^4 = 0$$

or $0.343 \, \omega^4 - 23308 \, \omega^2 + 1.51 \, E6 = 0$ \qquad (5.71)

Solving the above quadratic equation

$$\omega^2 = \frac{+23308 \pm 23264}{0.686}$$

$\omega_1^2 = 64.84;\quad \omega_1 = 8.05 \text{ rads/s};\quad n_1 = 1.28 \text{ Hz}$

$\omega_2^2 = 67889;\quad \omega_2 = 260.6 \text{ rads/s};\quad n_2 = 41.47 \text{ Hz}$

(c)   Zero axial force

In this case $[K_{Gii}]$ is null, so that the dynamical equation becomes:

$$\left| [K_{11}] - \omega^2 [M_{11}] \right| = 0$$

or $$\left| \begin{bmatrix} 5240 & 7860 \\ 7860 & 15720 \end{bmatrix} - \omega^2 \begin{bmatrix} 2.574 & 1.089 \\ 1.089 & 0.594 \end{bmatrix} \right| = 0 \qquad (5.72)$$

or $(5240 - 2.574 \, \omega^2)(15720 - 0.594 \, \omega^2)$

$$-(7860 - 1.089 \, \omega^2)^2 = 0$$

or $8.237 \, E7 - 3113 \, \omega^2 - 40463 \, \omega^2 + 1.529 \, \omega^4$

$$-6.178 \, E7 + 17119 \, \omega^2 - 1.186 \, \omega^4 = 0$$

or $\quad 0.343 \omega^4 - 26457 \omega^2 + 2.059 \, E7 = 0$ (5.73)

Solving the above quadratic equation,

$$\omega^2 = \frac{26457 \pm 25918}{0.686}$$

$\omega_1^2 = 786.3; \quad \omega_1 = 28 \text{ rads/s}; \qquad n_1 = 4.46 \text{ Hz}$

$\omega_2^2 = 76348; \quad \omega_2 = 276.3 \text{ rads/s}; \qquad n_2 = 43.98 \text{ Hz}$

NB   The Euler buckling load for this strut was 3232.3 N, and from the above results, it can be seen that as this load was approached, the fundamental natural frequency of vibration approached zero. It was also interesting to note that the fundamental mode was more sensitive to the axial compressive load, than the second eigenmode; the reason for this was that the fundamental mode of vibration was similar to the fundamental mode of buckling. This observation prompts the possibility that if a yacht mast were subjected to a periodic compressive axial load, it can fail through **dynamical instability** at a load a lot less than that required to cause static buckling.

Sec. 5.7] Non-Linear Vibrations 181

Examples for Practice 5

1. Determine the elastic instability loads for the plane pin-jointed trusses shown in Figures 5.8 to 5.10, given that AE = 5 MN.

   (a)

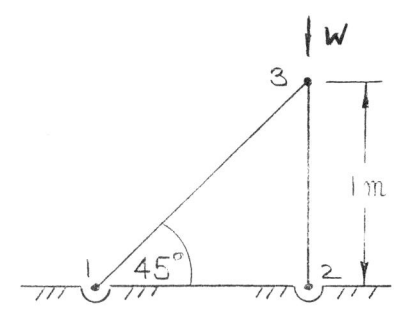

   Figure 5.8

   (b)

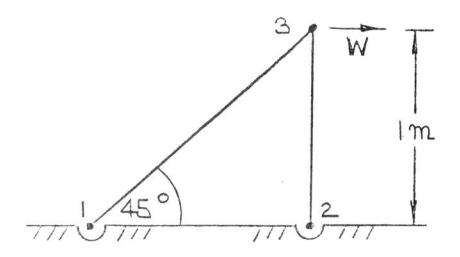

   Figure 5.9

   (c)

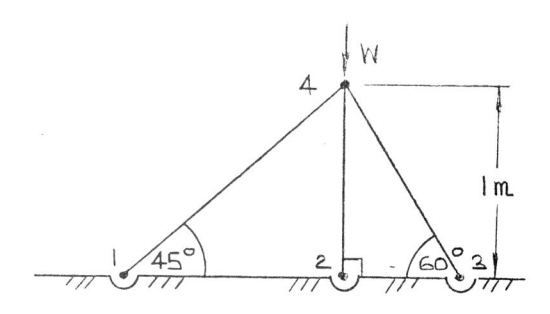

   Figure 5.10

2. Determine the elastic buckling load for a uniform section axially loaded strut, of length "ℓ", and fixed at both ends, but free to move axially inwards. (Hint: *use two equal length elements and use the property of symmetry*).

3. Determine the elastic buckling load for a uniform section, axially loaded strut, of length "ℓ", and fixed at one end and free at the other. (Hint: *use one element to mathematically model the strut*).

4. Determine the natural frequencies of vibration for the pin-jointed truss of Figure 5.11, when,

    (a)   W = 0

    (b)   W = 10 MN = compressive force; not a mass or a load.

    The following may be assumed to apply to this truss:

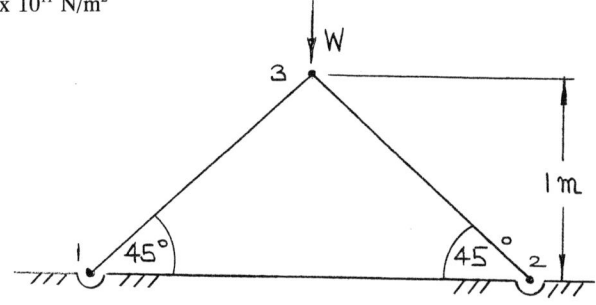

    A = 1.2 x 10⁻⁴ m²

    ρ = 7860 kg/m³

    E = 2 x 10¹¹ N/m²

Figure 5.11

Chapter 6

The Modal Method of Analysis

6.1 Introduction

The method of analysis described in this Chapter, is based on the finite element method, and it is suitable for determining the dynamic response of structures to time-dependent forces. Only simple structures will be considered in this Chapter, but the effects of damping will also be taken into account.

The method used here shows that for complex problems involving forced vibrations and damping, their solution can become extremely simple when the modal method of analysis is adopted.

6.2 The Modal Matrix $[\Phi]$

In this section, we will show how it is possible to uncouple the matrix dynamical equation of motion of an "n" degree of freedom forced vibration system, with the aid of the modal matrix $[\Phi]$. Now the modal matrix $[\Phi]$ is a square matrix, whose columns are the "n" eigenmodes of the structure, corresponding to the "n" degrees of freedom of the system, as shown by equation (6.1)

$$[\Phi] = [\Phi_1 \ \Phi_2 \ \Phi_3 \ \ldots \ \Phi_n], \qquad (6.1)$$

where

$$[\Phi_i] = \text{the } i^{th} \text{ eigenmode}$$

$$= \begin{Bmatrix} x_1 \\ x_2 \\ x_3 \\ x_j \\ x_n \end{Bmatrix} \qquad (6.2)$$

x_j = the jth relative displacement of the eigenmode "i"

Consider a two degree of freedom system, with the modal matrix $[\Phi]$ of equation (6.3)

$$[\Phi] = [\Phi_1 \ \Phi_2] \qquad (6.3)$$

Let $[M]$ = the mass matrix

and $[\overline{M}]$ = $[\Phi]^T [M] [\Phi]$ \qquad (6.4)

$$= \begin{bmatrix} \phi_1^T \\ \phi_2^T \end{bmatrix} [M] [\phi_1 \ \phi_2] \tag{6.5}$$

$$= \begin{bmatrix} \phi_1^T M \\ \phi_2^T M \end{bmatrix} [\phi_1 \ \phi_2]$$

or

$$[\overline{M}] = \begin{bmatrix} \phi_1^T M \phi_1 & \phi_1^T M \phi_2 \\ \phi_2^T M \phi_1 & \phi_2^T M \phi_2 \end{bmatrix}, \tag{6.6}$$

but as $[\Phi_1]$ and $[\Phi_2]$ are orthogonal,

$$[\Phi_1^T M \Phi_2] = [\Phi_2^T M \Phi_1] = [0] \tag{6.7}$$

Hence, equation (6.6) becomes of the diagonal form shown by equation (6.8)

$$[\overline{M}] = \begin{bmatrix} \phi_1^T M \phi_1 & 0 \\ 0 & \phi_2^T M \phi_2 \end{bmatrix} \tag{6.8}$$

$$= \begin{bmatrix} M_1 & 0 \\ 0 & M_2 \end{bmatrix}, \tag{6.9}$$

where

$$[M_1] = [\phi_1^T M \phi_1] \tag{6.10}$$

$$\text{and } [M_2] = [\phi_2^T M \phi_2] \tag{6.11}$$

Similarly, it can be shown that

$$[\bar{K}] = [\phi]^T [K] [\phi] \tag{6.12}$$

$$= \begin{bmatrix} K_1 & 0 \\ 0 & K_2 \end{bmatrix} \tag{6.13}$$

where

$$[K_1] = [\Phi_1^T K \Phi_1] \tag{6.14}$$

and $[K_2] = [\Phi_2^T K \Phi_2] \tag{6.15}$

NB Both $[\bar{M}]$ and $[\bar{K}]$ are of diagonal form, hence, it will be easier to handle them mathematically; this is taken advantage of by introducing the **weighted modal matrix** $[\bar{\Phi}]$ which is obtained by dividing $[\Phi]$ by the square root of $[M]$, as follows:

$$[\bar{\phi}] = [\phi] / \sqrt{[M]} \tag{6.16}$$

= weighted modal matrix

Now $[\bar{\phi}]^T [M] [\bar{\phi}]$

$= [\phi]^T [M^{-1}] [M] [\phi] = [I]$ = the identity matrix $\tag{6.17}$

and $[\bar{\phi}]^T [K] [\bar{\phi}]$

$= [\phi]^T [M^{-1}] [K] [\phi]$

but $[M^{-1}] [K] = [\lambda] \tag{6.18}$

$\therefore [\phi]^T [M^{-1}] [K] [\phi] = [\lambda]$

where

$$[\lambda] = \begin{bmatrix} \omega_1^2 & & & & \\ & \omega_2^2 & & & \\ & & \omega_3^2 & & \\ & & & \ldots & \\ & & & & \omega_n^2 \end{bmatrix} \tag{6.19}$$

ω_i = the ith radian frequency of vibration.

Now the matrix dynamical equation of motion of a structure, without damping, is given by:

$$[K] \{u_i\} + [M] \{\ddot{U}_i\} = \{q_t\} \tag{6.20}$$

where

$\{q_t\}$ = a vector of time-dependent forcing functions.

Let $\{U_i\} = [\bar{\Phi}] \{x_i\}$ (6.21)

and $\{\ddot{U}_i\} = [\bar{\Phi}] \{\ddot{x}_i\}$ (6.22)

Hence, equation (6.20) becomes

$$[K] [\bar{\Phi}] \{x_i\} + [M] [\bar{\Phi}] \{\ddot{x}_i\} = \{q_t\} \tag{6.23}$$

Pre-multiplying equation (6.23) by $[\bar{\Phi}]^T$, we get

$$[\bar{\Phi}]^T [K] [\bar{\Phi}] \{x_i\} + [\bar{\Phi}]^T [M] [\bar{\Phi}] \{\ddot{x}_i\} = [\bar{\Phi}]^T \{q_t\}, \tag{6.24}$$

or $[\Phi]^T [M^{-1}] [K] [\Phi] \{x_i\} + [\Phi]^T [M^{-1}] [M] [\Phi] \{\ddot{x}_i\} = [\bar{\Phi}]^T \{q_t\}$

$\therefore$ $[\lambda] \{x_i\} + [I] \{\ddot{x}_i\} = \{P_t\},$ (6.25)

where $\{P_t\} = [\bar{\Phi}]^T \{q_t\}$ (6.26)

$$[\lambda] = \begin{bmatrix} \omega_1^2 & & & & \\ & \omega_2^2 & & & \\ & & \omega_i^2 & & \\ & & & \ldots & \\ & & & & \omega_n^2 \end{bmatrix} \tag{6.27}$$

From equation (6.25), it can be seen that the equations are uncoupled and each row equation can be solved independently line by line.

Example 6.1

Determine expressions for the nodal displacements of the two element rod structure of Example 4.1, when it is subjected to periodical axial excitation force of value F(t), at its node 2. Hence, or otherwise, determine the $[\Phi]$ matrix, where $F(t) = F \sin \omega t$.

From Example (4.1),

$$[K] = [K_{11}] = 1E6 \begin{bmatrix} 180 & -80 \\ -80 & 80 \end{bmatrix} \tag{6.28}$$

$$[M] = [M_{11}] = \begin{bmatrix} 7.598 & 1.179 \\ 1.179 & 22.358 \end{bmatrix} \tag{6.29}$$

$$[\Phi] = [\Phi_1 \ \Phi_2],$$

where

$$[\phi_1] = \begin{bmatrix} 0.492 \\ 1.0 \end{bmatrix}$$

and $[\phi_2] = \begin{bmatrix} 1 \\ -0.215 \end{bmatrix}$

so that

$$[\phi] = \begin{bmatrix} 0.492 & 1.0 \\ 1.0 & -0.215 \end{bmatrix} \tag{6.30}$$

Now, the equation of motion is:

$$[K]\{U_i\} + [M]\{\ddot{U}_i\} = \{F(t)\} \tag{6.31}$$

Let $\{U_i\} = [\Phi]\{x_i\}$

and $\{\ddot{U}_i\} = [\Phi]\{\ddot{x}_i\},$ \hfill (6.32)

So that equation (6.31) becomes:

$$[K][\Phi]\{x_i\} + [M][\Phi]\{\ddot{x}_i\} = \{F(t)\} \qquad (6.33)$$

Substituting equations (6.28), (6.29) and (6.30) into equation (6.33), we get

$$\begin{bmatrix} 0.492 & 1.0 \\ 1.0 & -0.215 \end{bmatrix} 1E6 \begin{bmatrix} 180 & -80 \\ -80 & 80 \end{bmatrix} \begin{bmatrix} 0.492 & 1.0 \\ 1.0 & -0.215 \end{bmatrix} \{x_i\}$$

$$+ \begin{bmatrix} 0.492 & 1.0 \\ 1.0 & -0.215 \end{bmatrix} \begin{bmatrix} 7.598 & 1.179 \\ 1.179 & 22.358 \end{bmatrix} \begin{bmatrix} 0.492 & 1.0 \\ 1.0 & -0.215 \end{bmatrix} \{\ddot{x}_i\} = \begin{bmatrix} 0.492 & 1.0 \\ 1.0 & -0.215 \end{bmatrix} \begin{Bmatrix} 0 \\ F(t) \end{Bmatrix} \qquad (6.34)$$

Now

$$[\overline{K}] = [\phi]^T [K] [\phi] = \begin{bmatrix} 0.492 & 1.0 \\ 1.0 & -0.215 \end{bmatrix} 1E6 \begin{bmatrix} 180 & -80 \\ -80 & 80 \end{bmatrix} \begin{bmatrix} 0.492 & 1.0 \\ 1.0 & -0.215 \end{bmatrix}$$

$$= 1E6 \begin{bmatrix} 0.492 & 1.0 \\ 1.0 & -0.215 \end{bmatrix} \begin{bmatrix} 8.56 & 197.2 \\ 40.64 & -97.2 \end{bmatrix}$$

$$= 1E6 \begin{bmatrix} 44.85 & 0 \\ 0 & 218.1 \end{bmatrix} \qquad (6.35)$$

and $[\overline{M}] = [\phi]^T [M] [\phi] = \begin{bmatrix} 0.492 & 1.0 \\ 1.0 & -0.215 \end{bmatrix} \begin{bmatrix} 7.598 & 1.179 \\ 1.179 & 22.358 \end{bmatrix} \begin{bmatrix} 0.492 & 1.0 \\ 1.0 & -0.215 \end{bmatrix}$

$$= \begin{bmatrix} 0.492 & 1.0 \\ 1.0 & -0.215 \end{bmatrix} \begin{bmatrix} 4.917 & 7.345 \\ 22.938 & -3.628 \end{bmatrix}$$

$$= \begin{bmatrix} 25.36 & 0 \\ 0 & 8.125 \end{bmatrix} \tag{6.36}$$

and $\{P_t\} = [\phi]^T \{q_t\}$

$$= \begin{bmatrix} 0.492 & 1.0 \\ 1.0 & -0.215 \end{bmatrix} \begin{Bmatrix} 0 \\ F\sin(\omega t) \end{Bmatrix} \tag{6.37}$$

$$\therefore \{P_t\} = \begin{Bmatrix} F\sin(\omega t) \\ F\sin(\omega t) \end{Bmatrix} \tag{6.38}$$

The equation of motion becomes:

$$[K]\{x_i\} + [M]\{\ddot{x}_i\} = \begin{Bmatrix} F\sin(\omega t) \\ F\sin(\omega t) \end{Bmatrix}, \tag{6.39}$$

or

$$1E6 \begin{bmatrix} 44.85 & 0 \\ 0 & 218.1 \end{bmatrix} \{x_i\} + \begin{bmatrix} 25.36 & 0 \\ 0 & 8.125 \end{bmatrix} \{\ddot{x}_i\}$$

$$= \begin{Bmatrix} F \\ F \end{Bmatrix} \sin(\omega t) \tag{6.40}$$

Equation (6.40) can be seen to be uncoupled, with the following two independent equations:

$$44.85\ E6\ x_2 + 25.36\ \ddot{x}_2 = F \sin(\omega t) \tag{6.41}$$

and $\quad 218.1\ E6\ x_3 + 8.125\ \ddot{x}_3 = F \sin(\omega t) \tag{6.42}$

These two independent equations can both be seen to be of the form

$$k_i\ x_i + m\ \ddot{x} = F \sin(\omega t) \tag{6.43}$$

If we divide through by m_i, we get

$$\ddot{x}_i + \frac{k_i}{m_i} x_i = \frac{F}{m_i} \sin\omega t, \tag{6.44}$$

but $\quad \dfrac{k_i}{m_i} = \omega_i^2$

Hence, equation (6.44) becomes

$$\ddot{x}_i + \omega_i^2\ x_i = \frac{F}{m_i} \sin(\omega t) \tag{6.45}$$

The complementary function can be found by substituting $x_i = Ae^{\alpha t}$ in equation (6.45)

ie $\quad (\alpha^2 + \omega_i^2)\ x_i = 0$

or $\quad \alpha = \pm \omega_i$

$\therefore \quad x_i = A \cos \omega_i t + B \sin \omega_i t \tag{6.46}$

The particular integral is given by

$$x_i = \frac{F \sin(\omega t)}{m_i \left(D^2 + \omega_i^2\right)} = \frac{F \sin(\omega t)}{m_i \left(-\omega^2 + \omega_i^2\right)} \tag{6.47}$$

$$= \frac{F \sin(\omega t)}{m_i\ \omega_i^2 \left(1 - \omega^2/\omega_i^2\right)}$$

but $\omega_i^2 = k_i/m_i$,

where D = a differential operator

Hence the particular integral becomes

$$x_i = \frac{F \sin(\omega t)}{k_i \left(1 - \omega^2/\omega_i^2\right)} \tag{6.48}$$

From equations (6.46) and (6.48), the complete solution becomes:

$$x_i = A \cos \omega_i t + B \sin \omega_i t + \frac{F \sin(\omega t)}{k_i \left(1 - \omega^2/\omega_i^2\right)} \tag{6.49}$$

and

$$\dot{x}_i = -\omega_i A \sin \omega_i t + \omega_i B \cos \omega_i t + \omega \frac{F \cos(\omega t)}{k_i \left(1 - \omega^2/\omega_i^2\right)} \tag{6.50}$$

The boundary conditions are that at $t = 0$; $x_i = x_i(o)$ and $\dot{x}_i = \dot{x}_i(o)$.

Substituting the first boundary condition, into equation (6.49), we get

$$A = x_i(0) \tag{6.51}$$

Substituting the second boundary condition into (6.50), we get

$$\dot{x}_i(o) = \omega_i B + \frac{\omega F}{k_i \left(1 - \omega^2/\omega_i^2\right)}$$

or

$$B = \frac{\dot{x}_i(o)}{\omega_i} - \frac{\omega F}{\omega_i k_i \left(1 - \omega^2/\omega_i^2\right)} \tag{6.52}$$

Hence, from equations (6.51) and (6.52),

$$x_i = x_i(o) \cos \omega_i t + \frac{\dot{x}_i(o)}{\omega_i} \sin \omega_i t$$

$$+ \frac{F}{k_i \left(1 - \omega^2/\omega_i^2\right)} \left(\sin \omega t - \frac{\omega}{\omega_i} \sin \omega_i t\right) \tag{6.53}$$

The nodal displacements, u_2 and u_3 can be obtained from the expression

$$\begin{Bmatrix} u_2 \\ u_3 \end{Bmatrix} = [\phi] \begin{Bmatrix} x_2 \\ x_3 \end{Bmatrix} \tag{6.54}$$

$$= \begin{bmatrix} 0.492 & 1.0 \\ 1.0 & -0.215 \end{bmatrix} \begin{Bmatrix} x_2 \\ x_3 \end{Bmatrix} \tag{6.55}$$

The weighted modal matrix $[\bar{\Phi}]$ can be obtained from the expression:

$$[\bar{\Phi}] = \frac{[\phi]}{\sqrt{[\overline{M}]}} \tag{6.56}$$

where

$$\sqrt{[\overline{M}]} = \begin{bmatrix} \sqrt{25.36} & 0 \\ 0 & \sqrt{8.125} \end{bmatrix} \tag{6.57}$$

$$= \begin{bmatrix} 5.036 & 0 \\ 0 & 2.850 \end{bmatrix}$$

$$\therefore [\bar{\Phi}] = \begin{bmatrix} \dfrac{0.492}{5.036} & \dfrac{1.0}{2.850} \\ \dfrac{1.0}{5.036} & \dfrac{-0.215}{2.850} \end{bmatrix}$$

$$= \begin{bmatrix} 0.0977 & 0.351 \\ 0.1986 & -0.0754 \end{bmatrix} \tag{6.58}$$

6.3 Damping

The effects of damping on structural vibration is very important and in this section, we will consider three forms of damping, namely,

(a) viscous damping

(b) Rayleigh damping

(c) structural damping

6.3.1 Viscous Damping

The equation of motion for viscous damping is:

$$[K]\{U_i\} + [M]\{\ddot{U}_i\} + [c]\{\dot{U}_i\} = \{q_t\} \tag{6.59}$$

where, $[c] = \int_{vol} [N]^T [\mu] [N] \, d(vol)$

$[\mu]$ is a matrix of viscous damping terms.

It can be seen from equation (6.59), that viscous damping is proportional to a vector of velocities.

$$\text{Let } \{U_i\} = [\Phi]\{x_i\}$$

$$\text{So that } \{\dot{U}_i\} = [\Phi]\{\dot{x}_i\} \tag{6.60}$$

$$\text{and } \{\ddot{U}_i\} = [\Phi]\{\ddot{x}_i\}$$

Substituting equations (6.60) into equation (6.59), we get

$$[K][\Phi]\{x_i\} + [M][\Phi]\{\ddot{x}_i\} + [c][\Phi]\{\dot{x}_i\} = \{q_t\} \tag{6.61}$$

Pre-multiplying equation (6.61) by $[\Phi]^T$, we get

$$[\Phi]^T [K][\Phi]\{x_i\} + [\Phi]^T [M][\Phi]\{\ddot{x}_i\}$$
$$+ [\Phi]^T [c][\Phi]\{\dot{x}_i\} = [\Phi]^T \{q_t\}, \tag{6.62}$$

or

$$[\bar{K}]\{x_i\} + [\bar{M}]\{\ddot{x}_i\} + [\bar{c}]\{\dot{x}_i\} = \{P_t\} \tag{6.63}$$

where

$$[\bar{c}] = [\Phi]^T [c][\Phi] \tag{6.64}$$

Although $[\bar{K}]$ and $[\bar{M}]$ are of diagonal form, $[\bar{c}]$ is not, and because of this, equation (6.63) is uncoupled and difficult to solve.

6.3.2 Rayleigh Damping

If in equation (6.63), [c] is related to [K] and [M] in the following form:

$$[c] = \alpha [K] + \beta [M], \tag{6.65}$$

then equation (6.63) becomes uncoupled, where α and β are obtained experimentally.

This form of damping is called **Rayleigh damping** and it is often used in structural dynamics.

For Rayleigh damping,

$$[\overline{c}] = [\Phi]^T (\alpha[K] + \beta [M]) [\Phi], \tag{6.66}$$

so that,

$$[\overline{c}] = \alpha [\overline{K}] + \beta [\overline{M}] \tag{6.67}$$

Hence, equation 6.63) becomes

$$[\overline{K}] \{x_i\} + [\overline{M}] \{\ddot{x}_i\} + (\alpha [\overline{K}] + \beta [\overline{M}]) \{\dot{x}_i\} = \{P_{ti}\} \tag{6.68}$$

Equation (6.68) is uncoupled and it consists of a number of independent equations, each of the following form:

$$\ddot{x}_i + 2\zeta_i \omega_i \dot{x}_i + \omega_i^2 x_i = P_i \tag{6.69}$$

where

$$\zeta_i = \text{the modal damping ratio}$$
$$\text{and } 2\zeta_i \omega_i = \omega_i^2 \alpha + \beta \tag{6.70}$$

The two unknowns α and β can be obtained in terms of ζ_1, ζ_2, ω_1 and ω_2, from the first two equations of (6.70). From the first two equations of (6.70), we get,

$$\beta + \omega_1^2 \alpha = 2 \zeta_1 \omega_1 \tag{6.71}$$

$$\text{and } \beta + \omega_2^2 \alpha = 2 \zeta_2 \omega_2 \tag{6.72}$$

Solution of equations (6.71) and (6.72) gives

$$\alpha = 2(\omega_2 \zeta_2 - \omega_1 \zeta_1) / (\omega_2^2 - \omega_1^2) \tag{6.73}$$

$$\text{and } \beta = 2 \omega_1 \omega_2 (\omega_2 \zeta_1 - \omega_1 \zeta_2) / (\omega_2^2 - \omega_1^2) \tag{6.74}$$

and from equation (6.70), the damping ratios for the other modes is given by:

$$\zeta_i = \frac{\alpha \, \omega_i}{2} + \frac{\beta}{2\omega_i} \qquad (6.75)$$

The separate effects of mass-proportional and stiffness-proportional damping are shown in Figure 6.1.

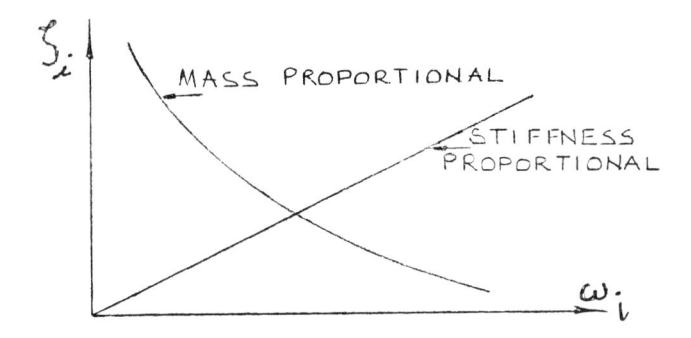

Figure 6.1 Mass-Proportional and Stiffness-Proportional Damping.

From Figure 6.1, it can be seen that the effects of mass-proportional damping on ζ_i, are larger for smaller values of ω_i and vice-versa for stiffness-proportional damping. Additionally, it can be seen that for stiffness-proportional damping, ζ_i is linearly proportional to ω_i.

To obtain the complimentary function of equation (6.70), it can be written in the form:

$$\ddot{x}_i + \frac{\mu}{m_i} \dot{x}_i + \frac{k_i}{m_i},$$

the solution of which, according to Case et al [3], is given by:

$$x_i = A e^{\left\{-\mu/2m_i + \sqrt{(\mu/2m_i)^2 - (k_i/m_i)}\right\} t}$$

$$+ B e^{\left\{-\mu/2m_i - \sqrt{(\mu/2m_i)^2 - (k_i/m_i)}\right\} t}$$

Now (k_i/m_i) is usually very much larger than $(\mu/2 m_i)^2$, hence, the expression for x_i can be simplified to the form

$$x_i = Ae^{(-\mu/2m_i + j\sqrt{k_i/m_i})t}$$

$$+ Be^{(-\mu/2m_i - j\sqrt{k_i/m_i})t}$$

$$= e^{-(\mu/2m_i)t}\left[Ae^{j\sqrt{k_i/M_i}t} + Be^{-j\sqrt{k_i/m_i}t}\right]$$

$$= e^{-(\mu/2m_i)t}\left[C\cos\left\{\sqrt{\frac{k_i}{m_i}}t + \epsilon\right\}\right]$$

Case et al [3] also show that if equation (6.70) is in the form:

$$m_i \ddot{x}_i + \mu \dot{x}_i + k_i = P \sin \omega t,$$

the particular integral is given by:

$$x_i = \frac{P \sin \omega t}{m_i D^2 + \mu D + k_i}$$

where
    D = a differential operator.

Case et al [3] give the following expression for the particular integral:

$$x_i = \frac{P\left[(k_i - \omega^2 m_i)\sin\omega t - \mu\omega\cos\omega t\right]}{(k_i - \omega^2 m_i)^2 + \mu^2 \omega^2}$$

$$= \frac{Pk_i\left[\left(1 - \frac{\omega^2}{\omega_i^2}\right)\sin\omega t - \mu\omega\cos\omega t\right]}{k_i^2\left(1 - \frac{\omega^2}{\omega_i^2}\right)^2 + \mu^2 \omega^2}$$

The maximum amplitude of this forced vibration, namely, $x_{i(max)}$, is given by:

$$x_{i(max)} = \frac{P}{\sqrt{k_i^2\left(1 - \omega^2/\omega_i^2\right)^2 + \mu^2 \omega^2}}$$

6.3.3 Structural Damping

For structural damping, Petyt [7] shows that the equation of motion is given by:

$$([K] + j[H]) \{U_i\} + [M] \{\ddot{U}_i\} = \{q_t\} \quad (6.76)$$

$$\text{where } j = \sqrt{-1}$$

Petyt explains that this form of damping can only be used when the excitations is harmonic, and that the complex stiffness matrix ([K] + j[H]) can be obtained by letting

the Complex Young's modulus $= E(1 + j\eta)$, $\quad (6.77)$

where E = Young's modulus of elasticity

and η = the material loss factor

= 2×10^{-5} for pure aluminium

= 1.0 for hard rubber.

Hence, from equation (6.77)

$$[H] = \eta[K], \quad (6.78)$$

So that equation (6.76) becomes

$$[K](1 + j\eta)\{U_i\} + [M]\{\ddot{U}_i\} = \{q_t\} \quad (6.79)$$

Letting

$\{U_i\} = [\Phi]\{x_i\}$ in equation (6.79), we get

$$[K](1 + j\eta)[\Phi]\{x_i\} + [M][\Phi]\{\ddot{x}_i\} = \{q_t\} \quad (6.80)$$

Pre-multiplying equation (6.80) by $[\Phi]^T$, we get

$$[\bar{K}](1 + j\eta)\{x_i\} + [\bar{M}]\{\ddot{x}_i\} = \{P_t\} \quad (6.81)$$

Equation (6.81) is uncoupled, and each horizontal line of equations can be solved quite independently of any other horizontal line of equations.

For further reading on this topic, see Petyt [7] and Thomson [8].

References

1. Ross C T F, Finite Element Methods in Engineering Science, Horwood, 1990.

2. Collar A R and Simpson A, Matrices and Engineering Dynamics, Horwood, 1987.

3. Case John, Lord Chilver and Ross Carl T F, Strength of Materials and Structures, Arnold, 1993.

4. Ross C T F, Mechanics of Solids, Prentice Hall, 1996.

5. Turner M J, Clough R W, Martin H C and Topp L J, Stiffness and Deflection Analysis of Complex Structures, J Aero.Sci, 23, 805-823, 1956.

6. Timoshenko S P and Woinowsky-Kreiger S, Theory of Plates and Shells, McGraw-Hill/Kogakusha, 1959.

7. Petyt M, Introduction to Finite Element Vibration Analysis, Cambridge University Press, 1990.

8. Thomson W T, Theory of Vibration, Unwin Hyman, 1989.

Answers to Further Problems

Examples for Practice 1

1(a) $\begin{bmatrix} 5 & 0 \\ 1 & 6 \end{bmatrix}$

1(b) $\begin{bmatrix} -1 & -2 \\ -3 & -2 \end{bmatrix}$

1(c) $\begin{bmatrix} 1 & 2 \\ 3 & 2 \end{bmatrix}$

1(d) $\begin{bmatrix} 4 & -2 \\ 1 & 7 \end{bmatrix}$

1(e) $\begin{bmatrix} 5 & -1 \\ 0 & 6 \end{bmatrix}$

2(a) 11

2(b) $\begin{bmatrix} 3 & 6 \\ 4 & 8 \end{bmatrix}$

3(a) $\begin{bmatrix} 0.6667 & 0.3333 \\ 0.3333 & 0.6667 \end{bmatrix}$

3(b) $\begin{bmatrix} 0.4 & -0.1 \\ -0.2 & 0.3 \end{bmatrix}$

4. $\lambda_1 = 2;$ $\quad [1 \quad -1]$
   $\lambda_2 = 5;$ $\quad [0.5 \quad 1]$

5(a) $\begin{bmatrix} 11 & 3 & 4 \\ 1 & 14 & 4 \\ 2 & 2 & 14 \end{bmatrix}$

5(b) $\begin{bmatrix} -1 & -1 & 2 \\ 3 & -2 & -4 \\ 6 & -4 & 0 \end{bmatrix}$

5(c) $\begin{bmatrix} 23 & 27 & 30 \\ 6 & 52 & 26 \\ 11 & 21 & 49 \end{bmatrix}$

Answers to Further Problems

5(d) $\begin{bmatrix} 38 & 17 & 25 \\ 27 & 43 & 25 \\ 24 & 9 & 43 \end{bmatrix}$

5(e) $\begin{bmatrix} 0.356 & -8.47\text{E}-2 & -0.153 \\ -0.119 & 0.195 & 5.08\text{E}-2 \\ -0.220 & 7.63\text{E}-2 & 0.237 \end{bmatrix}$

5(f) $\begin{bmatrix} 0.16 & -0.04 & 0.0 \\ -3.64\text{E}-3 & 0.16 & -9.09\text{E}-2 \\ 4.73\text{E}-2 & -8.0\text{E}-2 & 0.182 \end{bmatrix}$

5(g) λ_1 = 1.883;   [1   -0.469   - 0.872]
      λ_2 = 6.531;   [0.273   1   -0.194]
      λ_3 = 9.592;   [0.757   0.444   1]

5(h) λ_1 = 3.754;   [0.33   -0.865   1]
      λ_2 = 7.587;   [1   0.701   0.178]
      λ_3 = 9.656;   [0.712   1   0.592]

Examples for Practice 2

1(a) $u_4^o = 2.405/AE$; $v_4^o = 1.806/AE$; $F_{1-4} = 2.106$ kN; $F_{2-4} = 1.806$ kN;

$F_{3-4} = -0.590$ kN

1(b) $u_5^o = 10.527/AE$; $v_5^o = -4.21/AE$; $F_{1-5} = 3.789$ kN; $F_{2-5} = 2.105$ kN;

$F_{3-5} = -0.422$ kN; $F_{4-5} = -1.58$ kN

1(c) $u_2^o = 22/AE$; $v_2^o = -36.4/AE$; $u_3^o = -13.02/AE$; $F_{1-2} = 7.333$ kN;

$F_{1-3} = -4.17$ kN; $F_{2-3} = -6.01$ kN

2. $u_5^o = 32.98/AE$; $v_5^o = -32.31/AE$; $w_5^o = -24.74/AE$; $F_{1-5} = 1.29$ kN;

$F_{2-5} = -1.29$ kN; $F_{3-5} = -6.02$ kN; $F_{4-5} = -2.22$ kN

3(a) $v_2 = W\ell^3/3EI$; $\theta_2 = -W\ell^2/2EI$; $M_{1-2} = W\ell$; $M_{2-1} = 0$

3(b) $v_1 = -w\ell^4/8EI$; $\theta_1 = -w\ell^3/6EI$; $M_{1-2} = w\ell^2/2$; $M_{2-1} = 0$

4(a) $\theta_2 = 1.014 \times 10^{-2}/EI$; $\theta_3 = 7.56 \times 10^{-2}/EI$; $M_{1-2} = -0.268$ kN; $M_2 = \pm 0.308$ kNm;

$M_3 = \pm 0.823$ kNm; $M_{4-3} = 1.276$ kNm

4(b) $\theta_2 = -0.363/EI$; $v_3 = -0.829/EI$; $\theta_3 = 0.988/EI$ $M_{1-2} = -0.658$ kNm;

$M_2 = \pm 1.833$ kNm; $M_{3-2} = 0$

5. $u_2 = U_3^o = 22.4/EI$; $\theta_2 = 4.95/EI$; $\theta_3 = 1.45/EI$ $M_{1-2} = -8.59$ kNm;

$M_2 = \pm 0.784$ kNm; $M_3 = 5.62$ kNm; $M_4 = -9.01$ kNm

6. $u_2^o = 3.589 \times 10^{-8}$m; $v_2^o = -2.638 \times 10^{-9}$m; $\theta_2 = -7.922 \times 10^{-7}$ rads $M_{1-2} = -4.82$ kNm;

$M_2 = \pm 1.06$ kNm; $M_{3-2} = -0.527$ kNm

Examples for Practice 3

1. $[k] = \dfrac{AE}{\ell} \begin{bmatrix} 1 & -1 \\ -1 & 1 \end{bmatrix}$

2. $[k] = \begin{bmatrix} k_{11} & k_{12} & k_{13} \\ k_{21} & k_{22} & k_{23} \\ k_{31} & k_{32} & k_{33} \end{bmatrix}$,

where

$k_{11} = 3.333 AE/\ell;$ $\quad k_{12} = k_{21} = -4AE/\ell$

$k_{13} = k_{31} = 2AE/3\ell;$ $\quad k_{23} = k_{32} = -20AE/3\ell$

$k_{22} = 32AE/3\ell;$ $\quad k_{33} = 6AE/\ell$

3. $[k] = EI \begin{bmatrix} 12/\ell^3 & \text{symmetrical} & & \\ -6/\ell^2 & 4/\ell & & \\ -12/\ell^3 & 6/\ell^2 & 12/\ell^3 & \\ -6/\ell^2 & 2/\ell & 6/\ell^2 & 4/\ell \end{bmatrix}$

4. $[k] = GJ \begin{bmatrix} 1 & -1 \\ -1 & 1 \end{bmatrix}$

Examples for Practice 4

1. $[m] = \begin{bmatrix} m_{11} & m_{12} \\ m_{21} & m_{22} \end{bmatrix}$,

   where

   $m_{11} = CN\ (R_2^4/12 - R_1^4/4 - R_1^2 R_2^2/2 + 2 R_1^3 R_2/3)$

   $m_{12} = m_{21} = CN\ (R_2^4 - R_1^4 - 2 R_1 R_2^3 + 2 R_1^3 R_2)/12$

   $m_{22} = CN\ (R_2^4/4 - R_1^4/12 + R_1^2 R_2^2/2 - 2 R_1 R_2^3/3)$

   $CN = 2\pi\rho t/(R_2 - R_1)^2$

2. $\omega_1 = 1.732\ \ell \sqrt{(E/\rho)}$

3. $0;\ [\ 1\ \ 1\ ]$

   $\dfrac{3.464}{\ell} \sqrt{\dfrac{E}{\rho}}\ ;\ [1\ \ 1]$

4. $\omega_1 = \dfrac{3.533}{\ell^2} \sqrt{\dfrac{EI}{\rho A}}$

   $\omega_2 = \dfrac{34.81}{\ell^2} \sqrt{\dfrac{EI}{\rho A}}$

5(a) $n_1 = 486.4$ Hz; $[0.444\ \ 1.0]$; $\quad n_2 = 668.8$ Hz; $[1.0\ \ -\ 0.44]$

(b) $n_1 = 257.6$ Hz; $[1.0\ \ 0.132]$; $\quad n_2 = 457.1$ Hz; $[-0.132\ \ 1.0]$

6(a) $n_1 = 56.3$ Hz; $\quad n_2 = 73.6$ Hz; $\quad n_3 = 87.6$ Hz

(b) $n_1 = 42.4$ Hz; $\quad n_2 = 54.5$ Hz; $\quad n_3 = 77.2$ Hz

7(a) $n_1 = 308.9$ Hz; $\quad n_2 = 649.0$ Hz

(b) $n_1 = 251.5$ Hz; $\quad n_2 = 470.6$ Hz; $\quad n_3 = 925$ Hz

Examples for Practice 5

1(a)   1.306 MN

1(b)   3.536 MN

1(c)   7.48 MN

2. $40 \, EI/\ell^2$

3. $2.487 \, EI/\ell^2$

4(a) $n_1 = n_2 = 695.1$ Hz

4(b) $n_1 = n_2 = 583.7$ Hz

Index

acoustics 82
added mass 115
analyser 4
angle of twist 82
annular plate 95,96
axisymmetrically 95
axisymmetric shells 3
Azzi-Tsai 106

beam 26, 55, 57, 58, 62, 67, 85, 138, 142, 157,
    166, 167
    inclined 69, 148
    column 166, 167
bending moments 58, 61, 74, 76, 85-87
boundary element method 4
boundary values 55, 58, 89, 96, 100 110, 139

cantilever 142, 155
characteristic values 19
cofactor 12, 13
column 5
complex Young's modulus 197
composites 106
compound bar 31
constrained 34, 119
concentrated mass 151
continuous beams 26, 55, 61, 85, 138, 144, 157
co-ordinate transform 36

damping 193, 195
degrees of freedom 34, 88, 95, 99, 104, 138,
    148, 183
determinant 12, 13, 119, 124, 137, 147, 172
directional cosines 46, 47, 49, 54
distributed load 61, 104, 105
dynamical instability 180

eigenmodes 20-24, 120, 121, 125, 127, 135,
    152-157
eigenvalues 19-25
eigenvectors 25
elastic instability 160, 172
electrostatics 88
elemental stiffness matrix 29
encastré beam 58
end fixing
    moments 62, 66, 80
    forces 76-78, 105, 106
equivalent nodal loads 106
Euler buckling 159, 180

finite difference method 4
finite element method 4, 26, 29, 88, 104, 107,
    108, 175
fluid flow 88
four dimensional 128
frame 62, 67, 68, 76, 86, 87, 151
framework 26, 27

Gauss-Legendré 3
geometrical non-linearity 158, 159
geometrical stiffness matrix 158, 161, 164-166,
    168, 170, 171
global
    axes 35, 68
    axis 39, 41
    co-ordinates 34, 37, 38, 47, 114, 149,
        165, 168

Hermite polynomial 141
Hooke's Law 28, 43-45, 53, 54, 101
homogeneous equations 16, 19
hydrostatically loaded 86

in-plane 88, 95, 96, 98
isoparametric 3, 4

kinetic energy 108, 109

large deflections 158, 162, 166
latent roots 19
leading diagonal 6, 8, 18
local
    axes 35, 43, 53, 68
    axis 39, 41, 44
    co-ordinates 47, 68, 164, 168

magnetostatics 88
mass matrix 108, 110, 113, 114, 117, 126, 141,
    148-150, 183
material loss factor 197
material non-linearity 158
matrices
    addition of 9
    subtraction of 9
matrix
    band 8
    cofactor 14
    column 5
    diagonal 6, 16
    differentiation 11
    identity 7, 185

inverse 14, 16
integration 11
lower triangular 7
multiplication 10
null 8
reciprocal 14
row 6
scalar 6, 7
skew 8
skew symmetric 8
square 6
symmetric 8
trace of a 9
transpose of a 6
tri-diagonal 8
unit 7
upper triangular 7
matrix algebra 5
matrix displacement method 26, 85, 88
mid-side mode 99
minimum potential 94, 163, 167
minors 12
modal matrix 183, 185, 192
modal method 183

natural frequency 108, 116, 126, 127, 136, 142,
    151, 155-157, 175, 182
negative resultants 63, 77, 105
nodes 27
non-homogeneous equations 16, 17
non-linear 159, 168, 175
numerical instability 4

orthogonal matrix 36, 47
over-constrained 119

pin-jointed truss 26, 34, 38, 39, 83, 115, 121,
    127, 136, 155, 165, 169, 181
plane
    strain 93, 94
    stress 92, 93
plastic deformation 158, 159
plate elements 88
post-multiplier 10
post-processor 4
pre-multiplier 10
pre-processor 4
prescribed displacement 34
product 10

radian frequency 109

Rayleigh damping 193, 194
resonate 108
resultants 63
rigid-jointed 67, 70 76, 86, 148, 150, 151
rod 26, 27, 31, 34, 35, 37, 38, 45, 46, 99, 100,
    102, 113, 114, 116, 155, 161, 162, 165
rod inclined 69

scalar 5
settlement 34
shape functions 91, 96, 97, 101, 140
shell 88
simple harmonic motion 109
simultaneous equations 16, 17, 56, 74, 79, 80,
    88, 89, 96, 100, 139
slope-deflection 57, 60, 61, 66
space truss 27, 127, 136, 157
statically determinate 26
statically indeterminate 26
stiffness 26
stiffness matrix
    for a 3 node rod 102
    for a beam 57
    for a plate 95
    for a rod 26
    for a torque bar 83
strain energy 94, 162
stress contours 4
structural damping 193, 197
structural stiffness matrix 29, 30, 32, 65
superposition 62
system mass matrix 118
system stiffness matrix 32, 42, 51, 118, 126,
    171

tapered rod 107
three-dimensional trusses 45, 84
tie 166
time-dependent 186
torque bar 26, 82, 83
torsional constant 82
triangular form 18
tripod 49, 128, 156

vector 5
vibrations 88, 108, 115, 127, 136, 142, 144,
    151, 155, 168, 185
viscous damping 193
von Mises 106

Yacht mast 176